Arresting Development

T0316178

Scholars have become increasingly concerned about the impact of neo-liberalism on the field of development. Governments around the world have, for some time, been exposed to the forces of globalization and macro-economic reform, reflecting the power and influence of the world's principal international economic institutions and a broader commitment to the principles of neo-classical economics and free trade. Concerns have also been raised that neo-classical theory now dominates the ways in which scholars frame and ask their questions in the field of development.

This book is about the ways in which ideologies shape the construction of knowledge for development. A central theme concerns the impact of neo-liberalism on contemporary development theory and research. The book's main objectives are twofold. One is to understand the ways in which neo-liberalism and related world views of neo-classical theory and rational choice have framed and defined the 'meta-theoretical' aims and assumptions of what is deemed relevant, important and appropriate to the study of development. The second is to explore the theoretical and ideological terms on which an alternative to neo-classical theory may be theorized, idealized and pursued. By tracing the impact of Marxism, postmodernism and liberalism on the study of development, *Arresting Development* contends that development has become increasingly fragmented in terms of the theories and methodologies it uses to understand and explain complex and contextually specific processes of economic development and social change. Outside of neo-classical economics (and related fields of rational choice), the notion that social science can or should aim to develop general and predictive theories about development has become mired in a philosophical and political orientation that questions the ability of scholars to make universal or comparative statements about the nature of history, cultural diversity and progress.

To advance the debate, a case is made that development needs to re-capture what the American sociologist Peter Evans once called the 'comparative institutional method.' At the heart of this approach is an inductive methodology that searches for commonalities and connections to broader historical trends and problems while at the same time incorporating divergent and potentially competing views about the nature of history, culture and development. This book will be of interest to scholars and students of Development, Social and Political Studies and will also be beneficial to professionals interested in the challenge of constructing 'knowledge for development.'

Craig Johnson is Associate Professor of Political Science at the University of Guelph in Canada. He has published widely in the field of governance, decentralization and sustainable development, focussing mainly on Asia. His most recent publication (co-edited) is *Policy Windows and Livelihood Futures: Prospects for Poverty Reduction in Rural India* (Oxford University Press, 2006).

Arresting Development

The power of knowledge for social change

Craig Johnson

Routledge
Taylor & Francis Group

LONDON AND NEW YORK

First published 2009
by Routledge
2 Park Square, Milton Park, Abingdon, Oxon, OX14 4RN

Simultaneously published in the USA and Canada
by Routledge
711 Third Avenue, New York, NY 10017

Routledge is an imprint of the Taylor & Francis Group, an informa business

© 2009 Craig Johnson

Typeset in Times New Roman by Pindar NZ, Auckland, New Zealand

British Library Cataloguing in Publication Data
A catalogue record for this book is available from the British Library

Library of Congress Cataloguing in Publication Data
Johnson, Craig (Craig Anthony)
Arresting development : the power of knowledge for social change / Craig
Johnson.
    p. cm.
    Includes bibliographical references.
    ISBN 978-0-415-38154-3 (hbk) – ISBN 978-0-415-38153-6 (pbk) –
ISBN 978-0-203-08601-8 (ebk) 1. Economic development. 2. Social
change. 3. Knowledge management. I. Title.
    HD75.J64 2008
    338.9–dc22   2008025906

ISBN 13: 978-0-415-38154-3 (hbk)
ISBN 13: 978-0-415-38153-6 (pbk)
ISBN 13: 978-0-203-08601-8 (ebk)

ISBN 10: 0-415-38154-1 (hbk)
ISBN 10: 0-415-38153-3 (pbk)
ISBN 10: 0-203-08601-5 (ebk)

For Samuel, Lily and Nicholas

Contents

6 Advancing knowledge for social change   132

Boxes, tables and figures

Boxes

Tables

Figures

Preface and acknowledgements

This book is about whether, and on what basis, 'the past' may be used to inform efforts to change the future. The past I have in mind is one that is mired in 'postmodern' doubts about the ability of intellectuals and activists to represent history, and in a neo-classical scorn for anything that smacks of historical narrative. The future I have in mind concerns the coordinated efforts of governments, non-governmental organizations and social movements to improve the health, well-being and economic prosperity of peoples and societies around the world.

My inspiration for writing this book comes from a number of formative experiences, gained primarily during and after completing my doctoral studies in London in 2000. One, which happened during my 'viva' (the oral examination for the Ph.D.), was a question put to me by my examiner, David Mosse, about the 'a-historical' ways in which the theories I was using in my thesis considered the complexities of rural development and agrarian change. My answer was one of those responses that probably demonstrated a recognition that the question being put to me was important and relevant to my field of study but, in the back of my mind, I was left wondering about the ways in which history may be used to inform wider and predictive forms of knowledge. (In retrospect, I suspect that David was also left wondering whether I had fully answered the question, but (hopefully) on the strength of my other answers, he very kindly agreed to pass the thesis).

A second moment that would influence my interest in development theory was a brief and seemingly innocuous encounter with one of my first university colleagues, Robert Biel, from the Development Planning Unit at University College, London. Although the details are of course a blur to me now, one point that did stick in my mind was Robert's suggestion that development had become 'a field without theory.' I was intrigued by this, in part because the 'development' I knew at the time seemed to be full of 'theoretical' propositions about how and why communities, institutions, non-governmental organizations and social capital would make the world a better place. In retrospect, I realize that what he was talking about was the end of 'grand' social science theorizing about the historical and political conditions under which states and social movements would engage in political action to change the structure of society.

A final and truly formative moment in this journey happened during my

18 months of doctoral field research, which entailed what was essentially an effort to test theories of collective action and natural resource conservation by documenting the ways in which people experienced and acted upon local resource conflicts in a small fishing village in Southern Thailand (Johnson 2000, 2001, 2002). Although much could be said about the trials and tribulations I experienced during my time 'in the field,' one episode that stands out and captures, with some humour, the limitations of employing what was initially a positivist methodology, was the reaction of one particularly witty individual to my request that he fill in a household survey. Smiling, Bang Thee said that he would, but added quickly that the survey would be of far more use to him if he could use it for wrapping his fish!

Although I obviously didn't think about it at the time, Bang Thee's response to my request helped to capture what is for me a central theme of this book: of what use is social science research to the 'objects' of social science research? And on what basis may 'objects' of social science research be 'objectified' in social science research? For me, Bang Thee's joke raised questions about the ways in which key informants, focus groups and other data sources perceive the nature and importance of the questions being asked of them; to what extent and on what basis can (or should) we use the stated perceptions and preferences of individuals to inform a more general understanding of social phenomena and change? And finally, how may participatory research methodologies contribute to a theory of development that is historically aware of the power, politics and interests that affect the formulation of research questions and research agendas?

This book is about the ways in which competing ideologies have shaped the meta-theoretical aims and assumptions about what is deemed relevant, important and appropriate to the study of development.

Sections of Chapter Six and almost all of Chapter Two were previously published in C. Johnson (2004) *Uncommon Ground: The 'Poverty of History' in Common Property Discourse.* Permission to reproduce the article from Blackwell Publishers and from *Development and Change* is gratefully acknowledged.

I would also like to thank three anonymous reviewers as well as René Véron, Ryan O'Neill, Derek Hall, Theresa Lee, Dan Brockington, Bent Flyvbjerg, David Booth, Philip McMichael, Tim Shaw and, especially, Tim Forsyth for reading and providing comments on earlier sections of this book. For their words of inspiration and support I would like to thank Jordi Díez, Peter Vandergeest, Dennis Rodgers, Jane Parpart, Marta Rohatynskyj, Sally Humphries, Kerry Preibisch, Arjan de Haan, Fred Eidlin, Matt Fox, Ian Spears and David MacDonald. At Routledge, I would like to thank Andrew Mould, Zoe Kruze, Jennifer Page and Michael P. Jones for their support and assistance throughout the course of this project. I am also grateful to Alex Parisien for her research assistance and especially to Jennifer Harris, whose questions, clarifications and editing have added some much-needed clarity and coherence to the argument and text.

Finally, I would like to express a special word of thanks to my friends and family for supporting (and enduring) me through this process. First and foremost is Sara Moore, my wife, confidante and best friend, whose patience, love and support have been a constant source of strength and inspiration. I would also like to thank my parents, Ronald and Mavis Johnson, and Sara's parents, Bill and Sally Moore, for their love and support. Last, but not least, I want to thank my three wonderful, beautiful kids, Sam, Lily and Nicholas, for keeping their father in line (if not on time).

Craig Johnson
June, 2008

1 Deconstructing 'knowledge for development'

> Approaching development from a knowledge perspective reinforces some well-known lessons, such as the value of an open trade regime and of universal basic education. It also focuses our attention on needs that have sometimes been overlooked: scientific and technical training, local research and development, and the critical importance of institutions to facilitate the flow of information for effective markets.
>
> The World Bank (1998: 2)

Introduction

Towards the end of the last century, the World Bank unveiled an ambitious plan to transform itself from an institution whose corporate ethos had been defined primarily in terms of long-term development financing to one whose mission would now be about fostering 'knowledge for development.' In its annual *World Development Report* (World Bank, 1998), the Bank suggested that donors and national governments could narrow the 'knowledge gap' between rich and poor by helping low-income countries 'acquire, absorb and communicate' knowledge and information that might improve literacy, research capacity and the flow of information for markets and trade. The following year, and in collaboration with the United Nations and the governments of Japan, Germany and Switzerland, the World Bank announced the launch of a new 'Global Development Network,' based in New Delhi, whose primary goal was to 'enhance the quality and availability of policy-oriented research, strengthen the institutions that undertake this work and offer networking opportunities in order to address better the causes and possible solutions to poverty and meet the challenges of development' (Johnson and Stone, 2000: 3).

Whether it was leading change or merely keeping in step with what was already happening in the field, the World Bank's ideological makeover marked a powerful trend in development policy and practice. In 2006, the Canadian International Development Agency (CIDA) announced plans to foster 'policy linkages' between Canadian universities and Canada's foreign aid program. In so doing, it aimed to emulate the UK Department for International Development's (DFID) very long track record of supporting and drawing upon universities and other development research institutions, such as the Institute of Development Studies (IDS) at the

University of Sussex and the Overseas Development Institute (ODI) in London. In June 2007, IDS Director Lawrence Haddad announced to a gathering of development researchers in Ottawa plans to develop a 'learning laboratory' that would generate knowledge for development that is 'globally constructed and globally accessible.'

At first glance, the notion that DFID, the World Bank and other international development agencies may foster knowledge for development has strong intuitive appeal. For one, it suggests a more open and democratic arrangement in which scholars and their respondents could inform development policy. Second, it suggests that knowledge can foster the expansion of opportunities and 'capabilities' (Sen 1999 [2001]) for those struggling to thrive and survive in a rapidly globalizing economy. Third, and most importantly, it gives the impression that careful research and analysis – as opposed to power, influence and (especially) ideology – may decide the nature and content of public policy.

Upon closer inspection, the notion that research may be conducted in isolation from 'power, influence and ideology' dramatically underplays the challenge of fostering a knowledge that is 'globally constructed and globally accessible.' For one, the field of development is one that is funded, supported and consumed by agencies whose interest in development research is not only or necessarily academic in nature (Sylvester 1999; Fine 2001; Mosse 2005; Saith 2006). Critical reflections about the commercialization of development research (e.g. Harriss 2005; Chambers 2005; Bernstein 2005), highlight the ways in which donor priorities now influence the quality and diversity of what gets studied and what gets discussed in the context of development. Reflecting upon his 40-year career at the UK-based Institute of Development Studies, Robert Chambers (2005) has argued that the current emphasis on competitive bidding for funded research (primarily in the UK) has greatly diminished the ability of development researchers to ask and investigate questions whose relevance to policy discourses is weak or uncertain (cf. Mosse 2005). Others, such as John Harriss (2005) and Henry Bernstein (2005), have suggested that the close relationship that now exists between donors and development researchers has narrowed the intellectual and ideological range of issues and questions researchers can ask in the field (cf. Brohmann 1995; Harriss 2002; Bracking 2005; Hickey and Bracking 2005; Mosse 2005; Saith 2006).

Second, efforts to understand the Third World and/or development are mediated by the professional norms and practices (e.g. publication, promotion, tenure and prestige) that underlie the disciplines and the bureaucracies, non-governmental organizations and independent think tanks whose funding, research and ideology have become increasingly pronounced in development research (Cooper and Packard 1997; Fine 2001: especially, Chapters 8 and 11; Harriss 2005; Mosse 2005: Chapter 6). Although development researchers have, in recent years, aimed to improve the ability of poor people to engage in the research process (see Chapter Five), development research (like any form of research) is rooted in a culture whose norms and practices are not necessarily receptive to the long-term needs and perspectives of the poor (Spivak 1988; Kapoor 2002, 2004; Parfitt 2002).

Third, the field is, in many instances, also rooted in a deeper historical context of colonialism and post-colonial reflection about the meaning and nature of colonial rule (e.g. Said 1979 [1994]; Spivak 1988; Goss 1996; Sylvester 1999; Kapoor 2004). Whether scholars make these relations explicit in their research, such histories affect the values and meaning social science researchers attach to the questions and topics they pursue, and they affect the relationship between the object and subject of social science research (Spivak 1988; Goss 1996; Sylvester 1999; Kapoor 2004). Spivak (1988), for instance, argues that efforts to represent Third World experiences constitute a form of intellectual imperialism in which meaning and data are extracted from respondents whose relationship with the research process will always be marginal.

Finally, there is the notion that development operates in a wider political economy, which frames and constrains the realm of what is askable, what is knowable and what is possible (Cowen and Shenton 1996; Gore 2000; Thomas 2000; Fine 2001). Michael Cowen and Robert Shenton (1996), for instance, have argued that the very idea of 'development' is a social construction whose explicit aim was to legitimate, accommodate and control 'the social disruptions caused by the unchecked "development" of capitalism' (Thomas 2000: 2). Similarly, and more recently, it has been argued that development is part of a wider and primarily Western agenda whose 'discursive' practices and interventions subject poor and politically marginal groups and nations to the interests, needs and ideologies of large development bureaucracies, such as the International Monetary Fund (IMF), the United Nations and the World Bank.[1]

In short, the idea that research and knowledge may exist in isolation from power, influence and ideology dramatically underplays the idea that knowledge is framed and constrained by a wider set of institutional, ideological and material factors that have a bearing on the kinds of questions and facts that are asked and pursued in the context of social science research. Similarly, the notion that international development agencies, such as the World Bank, may foster an objective body of 'knowledge for development' dramatically underplays the kinds of aims and values that shape their understanding of what is useful and what is important in the context of development. Indeed, the quotation that appears at the beginning of this chapter suggests that the World Bank's understanding of knowledge for development is one that is primarily interested in 'reinforcing' a particular kind of development, that is, one that appreciates 'the value of an open trade regime' (World Bank 1998: 2).

This book is about the ways in which ideologies shape the construction of knowledge for development. A central point of departure is the now dominant role of neo-liberalism and neo-classical theory in contemporary development theory and research. The book's main objectives are twofold. One is to understand the ways in which neo-liberalism and related worldviews of neo-classical theory and rational choice have framed and defined the 'meta-theoretical' aims and assumptions of what is deemed relevant, important and appropriate to the study of development. A second is to explore the theoretical and ideological terms on which an alternative to neo-classical theory may be theorized, idealized and pursued.

For the purposes of this analysis, *theory* represents 'a coherent body of generalizations and principles associated with the practice of a field of inquiry' (Chilcote 1994: 367). *Meta-theory* is 'theory about theory, where 'meta' refers to that which is 'beyond' theory or, more precisely, that which lies beyond the theory's pre-suppositions,' (Morrow and Brown 1994: 46). Meta-theoretical questions about ontology, epistemology and the construction of knowledge are centrally concerned with the foundational and ethical assumptions we make about what is deemed relevant, important and appropriate to the study of social phenomena. They pre-determine not only the concepts we use to study the world around us, but also the ethical considerations about what we should do with our research findings and why we need them in the first place. Finally, *ideology* represents a 'coherent and comprehensive set of ideas that explains and evaluates social conditions, helps people understand their place in society and provides a program for social and political action,' (Ball and Dagger 2004: 4).

Box 1.1　Theory, meta-theory and ideology: some working definitions

Neo-liberalism in theory and practice

Over the last three decades, development scholars have become increasingly concerned about the impact of neo-liberalism on the field of development – and with good reason. For some time, governments around the world have been exposed to the forces of globalization and macro-economic reform, reflecting the power and influence of the world's principal international economic institutions (i.e. the World Bank, the World Trade Organization and the IMF) and a broader commitment to the principles of neo-classical economics and free trade. Within this context, concerns have been raised that neo-classical theory now dominates the ways in which scholars frame and ask their questions in the context of development. In a special issue of *World Development*, for instance, Ravi Kanbur (2002: 477) has argued that:

> Development economics (by which he means neo-classical economics) stands in beleaguered ascendancy, atop development studies and development policy. Economists and economic thinking dominate the leading development institutions. The prestige of development economists within academia ... has never been stronger.

Similar concerns have been raised by Toye (1993), Brohmann (1995), Fine (1999; 2001), Mohan and Stokke (2000), Rapley (2002), Harriss (2001; 2002), Bracking (2004), Bevan (2004), Hickey and Mohan (2005), Kanbur and Shaffer (2006), Saith (2006), Eyben (2006) and Sumner (2007).

At the heart of neo-classical theory are four basic assumptions about the nature of social relations and economic life.[2] One is the assumption that social outcomes may be explained primarily in terms of the 'maximizing' calculations that individuals make about the perceived costs and benefits of future actions (methodological individualism). In Elinor Ostrom's words, the assumption here is that 'individuals compare expected benefits and costs of actions prior to adopting strategies for action' (Ostrom 1991: 243).

A second assumption is that individuals possess 'stable preferences' for the basic 'aspects of life,' including 'food, honor, prestige, health, benevolence, and especially wealth' (Gilpin 2001: 51).[3] Such preferences, it is argued, are universal, and apply irrespective of status, culture, history or wealth.

A third assumption, and one that is central to neo-classical economics, is the idea that markets emerge 'naturally' to coordinate 'with varying degrees of efficiency' (Gilpin 2001: 51) the actions and preferences of utility maximizing individuals.[4] Although there is diversity in the field, the general idea is that value reflects the utility goods and services may provide in relation to each additional – or marginal – unit of consumption. In the absence of market distortions (e.g. subsidies, collusion, etc.), it is argued, price mechanisms will provide the most efficient means of regulating supply and demand, producing a natural state of equilibrium whereby supply and demand will adjust to price variations (i.e. as demand increases in relation to supply so too will prices, increasing supply and/or reducing demand, putting a downward pressure on prices and so on).

A final assumption, and one that informs both neo-classical economics and rational choice theory (see Chapter Two), is the idea that hypothetic-deductive models of individual decision making may form the basis for comparison, generalization and the construction of theory. As Paul Pierson (2004: 168) has argued,

> The building blocks for theory are maximizing individuals, typically treated in a highly atomized way and portrayed as possessing core traits that are largely separable from any particular context. Theorizing grounded in rational choice analysis typically has an ambitious agenda of establishing claims that should apply, at least on average, across a wide range of settings whenever a few crucial conditions hold.

Therefore, neo-classical theory embodies the idea that social relations and outcomes may be explained primarily in terms of individualized actions and decisions. Neo-liberalism extends this logic by supporting policies and worldviews that would support and expand the ability of individuals to make basic decisions concerning the investment of time and money and the consumption of goods and services. In many developing countries, neo-liberalism has also entailed a wide set of programs and policies aimed at dismantling or considerably reducing the ability of national governments to protect vulnerable groups and sectors of the economy from market volatility and economic downturn.

The following section now provides a broad overview of the ways in which neo-liberal and neo-classical thinking has affected development theory, policy and research.

The debt crisis and the 'Washington consensus'

Rooted in neo-classical assumptions about marginal utility, market equilibrium and rational choice, neo-liberal 'solutions' (Rapley 2002) to the international debt crisis of the 1970s (see Box 1.2) used the apparent failure of central planning and import substitution to justify policies that would remove or relax the vestiges of state intervention and market regulation (Rapley 2002; McMichael 2004). Principal among these were fiscal and trade policy reforms that would (*inter alia*) reduce fiscal deficits, liberalize trade and investment and de-regulate domestic markets (Gore 2000; Rapley 2002; McMichael 2004; Craig and Porter 2006).

Central to the 'Washington consensus' that began to emerge during the 1980s were three inter-related assertions about the ability of states to generate and allocate resources within society. One was the notion that central state agencies lack what the Austrian economist Friedrich von Hayek once called 'time and place knowledge' about people's real needs and preferences (Hayek, cited in Ostrom *et al.* 1993: 51; Hayek, 1944 [1994]). More efficient, Hayek argued, were markets based on price mechanisms, private property and (relatively) unfettered trade (Hayek 1944 [1994: 55–56]). A second and related argument was that public ownership, price controls and other forms of state intervention 'crowded out' private investment, reducing economic productivity and inflating consumer prices (Rapley 2002). A third was that unchecked authority and inadequate incentives (reflected in salaries, rules of promotion and so on) created bloated bureaucracies and corrupt practices that diverted scarce resources from otherwise productive forms of investment (Krueger 1998; Lal 1983 [2000]).

In the long term, it was argued, structural adjustment would improve access to international markets, foreign exchange earnings and, more generally, it would foster a better life for those at the lower end of the socio-economic ladder (by reducing net consumer prices and stimulating agricultural and commodity markets) (Lal 1983 [2000]; Greenaway 1998; Krueger 1998). However, in the short term, the austerity measures that were used to restore financial stability entailed a number of shocks that were, in retrospect, deeply destructive to the lives and long-term sustainability of many low-income countries (Rapley 2002; Stiglitz 2003; McMichael 2004). First, the immediate emphasis on currency devaluation decimated real incomes, pushing many families and economies deeper into poverty. Second, the emphasis on deregulation, fiscal austerity and public sector reform meant massive cuts in state spending, which reduced or entirely removed important subsidies to agriculture, industry and public sector employment. Third, the deregulation of national and international commodity agreements exposed many developing countries to the volatility of world commodity prices (Page and Hewitt 2001; Rapley 2002). Fourth, the liberalization of national trade restrictions (i.e. tariffs, duties, quotas and licenses) exposed many Third World economies to competition from foreign imports, especially in agriculture. Fifth, the emphasis

The debt crisis of the 1980s can be traced directly to a series of decisions and events that in the early 1970s began to undermine the relative stability of dollar-gold parity and fixed exchange rates that had been the institutional pillars of the Bretton Woods era (Rapley 2002; McMichael 2004; Wade 2006). One was the Nixon Administration's decision in 1971 to abolish the gold standard, which linked the value of the US dollar to the price of gold. A second was the combined impact of the oil shocks of 1973 and 1979. The end of the gold standard reflected a number of structural factors, especially including the re-emergence of the European and Northeast Asian economies, the rise of offshore dollar-based money markets and the declining state of America's trade and fiscal balance. The first 'oil shock' of 1973 stemmed from the decision of the oil exporting economies of the Middle East to punish American support for Israel during the Yom Kippur War by withholding oil exports, creating a major recession in the West. Between 1973 and 1978, Western banks that were now flush with 'petro-dollars' (dollars earned from increased oil revenues) began extending credit to borrowers around the world, especially ones with large and growing public sectors (i.e. ones in the developing world). Finally, in 1979, a second oil shock drove the price of oil from \$13 to \$30 a barrel, creating another Western recession fuelled by 'hyper-inflation.' As central banks raised interest rates to control inflation, Western credit became increasingly expensive for international borrowers. By 1982, Mexico, Argentina and Brazil announced that they could no longer meet their sovereign debt obligations, ushering in the era of structural adjustment and neo-liberal reform.

Source: Rapley (2002); McMichael (2004)

Box 1.2 The oil shocks and the debt crisis

on exports shifted agricultural production from food crops to cash crops, creating new vulnerabilities to micro- and macro-economic fluctuations in prices, supply and demand. Finally, the liberalization of capital accounts (i.e. the liberalization of currency exchanges) exposed many emerging economies (especially in Latin America, East Asia, Eastern Europe and the Former Soviet Union) to the risk of currency speculation and capital flight (Rapley 2002; Wade 2006).

Neo-liberalism with a human face? 'The post-Washington consensus'

In the wake of the Russian and East Asian financial crises (both culminating in 1998), a new consensus of opinion began to question the notion that goods and services could be distributed only or primarily on the basis of supply and demand

(Stiglitz 2003; Craig and Porter 2006). First, the idea that markets would emerge and evolve purely on the basis of supply and demand significantly underplayed the conditions under which markets failed to provide a means of connecting utility-maximizing individuals (North 1990; Rapley 2002; Stiglitz 2003; Craig and Porter 2006). Second, the idea that social phenomena could be explained primarily – or only – on the basis of individual decision making underplayed the impact of structure and power on the rationality that informs individual decision making and choice (North 1990; Evans 1995: Chapter 2; Gilpin 2001: Chapter 3; Stiglitz Chapter 3; Sen 1999 [2001] ; Craig and Porter 2006).

One important response to the neo-classical orthodoxy was the new institutional economics (NIE). Like neo-classical economics, the NIE assumed that preferences are stable and that individuals seek to maximize personal utility, but that their ability to do so is *bounded* by a finite capacity to obtain, synthesize and utilize information. Unlike neo-classical economics, the new institutionalism posits that variations in markets and historical economic transformations reflect the impact of institutions, what Douglass North famously called the 'humanly devised constraints that structure human interaction' (North 1995: 23). Framed in this way, property rights, labour contracts, firms, prices and insurance all affect the *transaction costs* of searching for, synthesizing and monitoring relevant actors and events, obtaining information about price, quality, potential buyers, etc. and ensuring that other parties acted in good faith.[5]

Reformulating the idea that good governance was essentially a matter of 'getting the prices right,' scholars and practitioners of development (the World Bank in particular) therefore began to embrace the idea that institutions provided an important means of correcting or affecting market imperfections, fostering what many began to call a 'post-Washington consensus' (Mohan and Stokke 2000; Rapley 2002; Onis and Senses 2005). As DiJohn and Putzel (2000: 1) have argued:

> Emphasis by international development agencies is increasingly placed on the protection and enforcement of property rights and contracts, reliable legal structures, and other regulatory structures like competition policy and bankruptcy law ... The idea that market economies depend crucially on effective state institutions and state capability to create and enforce growth-enhancing 'rules of the game' or 'incentives' underlies the current interest in 'good governance.'
>
> (DiJohn and Putzel 2000: 1)

'What was needed,' Craig and Porter (2006: 12–13) have argued,

> ... was not shaky and corrupt government further undermined by anti-statist reforms, but an *enabling* state that could support the crucial institutions of law, financial and policy transparency and market information that the New Institutional Economics (NIE) held were basic to the emergence of efficient and competitive markets.

However, the notion that mainstream thinking (both within and beyond the World Bank) has reached 'a consensus' on the need to 'get institutions right' is difficult to sustain, particularly if this entails a rejection of free market principles and economic growth (Gore 2000; Onis and Senses 2005; Saith 2006; Craig and Porter 2006). First, cuts in state spending have had (and continue to have) a serious long-term impact on state capacity in the developing world (Rapley 2002; McMichael 2004; Onis and Senses 2005; Craig and Porter 2006). Second, the ideas that were most central to the Washington Consensus (liberalization, privatization, fiscal austerity and public sector reform) are still as much a part of the core principles on which the IMF and the World Bank design and evaluate their programs as they were during the days of structural adjustment (Fine 2001; Rapley 2002; Wade 2006; Craig and Porter 2006). Finally, the notion that the post-Washington consensus has entailed a rejection of free market principles and liberalized trade understates dramatically the extent to which liberalization and globalization are now embedded in the aims and assumptions of what is now characterized as development (Craig and Porter 2006).

In short, the debt crisis marked and facilitated a profound shift in thinking about the means by which scholars and practitioners of development (states, non-governmental organisations [NGOs], firms) should understand and address the political economy of development (Gore 2000; Rapley 2002). Beyond the assumption that the benefits of liberalized trade (and the economic growth this would produce) would eventually 'trickle down' to the poor (e.g. Krueger 1998; Greenaway 1998) was a more important and far-reaching assertion that the best way of achieving growth and reducing poverty was to structure institutions in such a way that individuals had the strongest incentives possible to economize, and to engage in 'market behaviour.'

Neo-classical theory: a 'colonizing' concept?

For scholars wedded to the idea of connecting theory (and policy) with an understanding and appreciation of history and context, the perceived dominance of neo-liberalism and neo-classical theory is deeply unsettling in a number of ways. First, it assumes a world in which individual values and preferences may be measured on the basis of statements made in the context of experimental research design, or choices made in the context of a (largely theoretical) market exchange (Brohmann 1995; Bracking 2004; Harriss 2002; Bevan 2004; Kanbur 2002; Kanbur and Shaffer 2006). Second, it appears to take little if any interest in history (Pierson 2004; Fine 2001; Harriss 2002; Bracking 2004; Brohman 1995; Bevan 2004). Third, it provides little if any room for definitions or interpretations outside of a seemingly artificial experimental research design (i.e. self-definitions of poverty, exclusion, etc. have no place in this analysis) (Bevan 2004). Finally, it offers what is for many an under-socialized model of individual decision making and rational choice. In Rosalind Eyben's words, 'societal processes and outcomes are seen as the sum of discrete, intentional acts by autonomous actors who are pre-constituted rather than defined through relations with others' (Eyben 2006: 603; cf. Gore 2000).

To appreciate the enduring influence of neo-classical theory on development policy, it is worth reflecting briefly upon the publication of two influential reports, both by the World Bank. One was the so-called 'East Asian miracle report' which, in 1993, argued that the principal factors explaining economic growth in the high-performing economies of South Korea, Taiwan, Indonesia, Malaysia and Thailand were a commitment on the part of national governments to principles of macro-economic stability and liberalized trade. To scholars familiar with the historical record of economic growth in the region, the World Bank's account of what led to the miracle of East Asian development appeared to ignore the crucial ways in which national governments (especially, but not only, in Korea and Taiwan) had manipulated and sheltered domestic markets through tariffs, quotas, subsidies and directed credit (see, for instance, Wade 1990 and Amsden 1989). For many, the views being expressed in the Asian miracle report said less about the quality of research methodology than it did about the ideological commitment of the World Bank to the principles of free market capitalism and neo-liberal reform. In the words of one writer (Wade 1996), the East Asian miracle report was essentially an effort to 'maintain' a paradigm rooted in the principles of neo-classical theory and neo-liberal reform.

A second and possibly more revealing moment in current history was the publication of the World Bank's annual *World Development Report: Attacking Poverty*, in 2000. In this case, the lead author was an economist (Ravi Kanbur) whose commitment to the neo-classical orthodoxy was not as strong as that of the mainstream and whose approach to the preparation of the report entailed an open commitment to alternative perspectives on development theory and policy. Influenced by the participatory ethos of the day, Kanbur invited scholars (primarily from the United Kingdom and the United States) to prepare papers (on the politics of poverty, among other things) that would (in theory) inform the empirical and normative content of the final *World Development Report*. Complementing this largely academic discussion was a series of 'Consultations with the Poor' – a participatory exercise that involved surveys and interviews with an estimated 60,000 people in 60 countries around the world (Wade 2001: 1436).

From an early stage in the process, discussions about the nature and content of the 2000 *World Development Report* concerned the relative merits of promoting policies that would protect people adversely affected by market volatility and economic downturn (Wade 2001). Broadly speaking, scholars were divided about the desirability and viability of using social safety nets to improve the security of poor and vulnerable groups in society. Some, such as the Yale economist T. N. Srinivasan, argued that a 'return' to social protection would reduce the prospects for economic growth that were necessary for poverty reduction (Wade 2001). Others, such as then US Treasury Secretary Larry Summers, argued that the emphasis on social welfare and redistribution would detract from the 'meta-theoretical' goals of market liberalization and free trade (Wade 2001). Caught in the middle of this ideological struggle was Kanbur, whose commitment to the participatory ethos of the report suggested that the studies and consultations that preceded the report were articulating views that differed from the neo-classical orthodoxy.

In the end, Summers and the US Treasury got their way; the final version of the *World Development Report* removed or significantly reduced any reference (made in earlier drafts of the report) to the notion that social safety nets and currency controls may be used to manage or restrict the flow of international trade and investment (Wade 2001). References to vulnerability and inequality were now mentioned only in relation to the perceived merits (indeed, the necessity) of trade liberalization and economic growth. And in May 2000, 5 months before the final version was released, Ravi Kanbur resigned as director of the *World Development Report*.

Box 1.3 Neo-liberalism in practice: wars of words at the World Bank

A wider concern is that neo-classical assumptions about individual decision making and rational choice now dominate the ways in which development is theorized, and pursued. Ben Fine (2001), for instance, has argued that the World Bank's use of the idea of 'social capital' reflects a more radical retreat from the structuralist/ historicist theorizing of Pierre Bourdieu (1977) and favours instead what is essentially a neo-classical rendering of the ways in which utility-maximizing individuals generate and exploit 'networks of trust' and 'norms of reciprocity' (Putnam 1993). The problem, Fine argues, is that efforts to model and measure trust, civil society and other aspects of social capital undervalue or ignore entirely wider structural and historical factors including, *inter alia*, 'the role of the nation-state, the exercise of power and the divisions and conflicts that are endemic to capitalist society' (Fine 2001: 191).

Similarly, Mohan and Stokke (2000) have argued that neo-liberalism and with it post-structuralism (see below) have produced an orientation that significantly underplays the impact of grand historical factors and processes, embracing instead a 'localism' that aims to foster development through decentralization, participation, 'social capital' and a decidedly non-radical rendering of social mobilization.

The problem, they argue, is twofold: first, it essentializes and romanticizes 'the local,' suggesting that development can be conceptualized and addressed through micro-economic models of community-based development; second, its view of the local is framed 'in isolation from broader economic and political structures,' suggesting that development processes can be conceptualized and decided through individualized models of rational choice (Mohan and Stokke 2000: 249).

A related concern is that neo-classical renderings of development theories and concepts now frame and therefore constrain what is possible and desirable in the context of development policy and practice. Ashwani Saith (2006), for instance, has recently argued that the Millennium Development Goals (MDGs) foster a culture that has become pre-occupied with short-sighted goals and targets, as opposed to long-term systematic change. Although the goals of cutting world poverty in half, achieving universal primary education, promoting gender equality, etc. are not on their own unworthy of our attention, the concern is that the MDGs are framed so broadly and ambiguously that they fail to articulate with sufficient clarity or purpose the ways in which governments, NGOs and people may act to address the underlying factors and processes that perpetuate poverty, hunger, inequality, etc. (Saith 2006; Eyben 2006). As Saith (2006) has argued, they are also conspicuously silent about the historical and causal relationship between the poverty reduction agenda and the (neo-liberal) conditions under which poverty reduction strategies are now being pursued.

Therefore, an over-riding concern is that neo-classical theory has infiltrated, or in Fine's terms 'colonized,' the field of development, framing and defining discourses concerning the need for neo-liberal reform, but also framing and defining the meta-theoretical aims and assumptions of what is deemed relevant, important and appropriate to the study of development.

Re-politicizing development: the elusive quest for unified theory

Towards the end of his analysis, Fine raises the crucial point that neo-classical renderings of social capital, and with it neo-liberalism, are not inevitable. They 'can be prevented,' he argues, and

> Doing so depends upon pursuing alternatives based upon scholarly integrity, genuine interdisciplinarity, and the resuscitation of political economy within and across the social sciences.
>
> (Fine 2001: 200)

But to what extent is there 'an alternative?' And to what extent can it be pursued in the current historical, intellectual and ideological climate?

Fine's call for an 'alternative' social science resonates strongly with assertions made by many scholars that development – and social science more generally – has lost its sense of politics, and of history. Mick Moore and James Putzel (1999), for instance, have argued that development scholars and agencies need to be more explicit about politics and power, by which they imply a wide range of 'political issues,' such as corruption, redistributive land reform, social mobilization and democratic reform. Similarly, John Harriss (2000; 2001) has argued that development needs to be understood in relation to the political opportunities that states and regimes make available, and the ways in which poor and marginal groups are able to exploit these opportunities. Finally, and more squarely directed towards the international aid regime, Rosalind Eyben has argued that development needs to move beyond the neo-classical emphasis on goals and targets, and embrace a paradigm that is 'more open to different ways of thinking about economy, society and politics' and therefore 'better able to support transformative processes' of social change (Eyben 2006: 595).

Fine's call for 'genuine interdisciplinarity' also echoes assertions made by many that development needs to embrace an epistemic commitment to the ideals of interdisciplinary research (Fine 2001; Harriss 2002; Kanbur 2002; Kanbur and Shaffer 2006). Harriss, for instance, has argued that 'good scholarship' requires a 'tension' between the rules and associated practices of the established academic disciplines (especially economics) and 'a healthy disrespect for the rules when they stand in the way of the pursuit of knowledge' (Harriss 2002: 488). Similarly, Kanbur and Shaffer (2006) argue that 'combined approaches,' involving qualitative and quantitative methodologies – what they call 'Q-squared' – may provide a way of bridging the positivist assumptions of neo-classical economics and the 'hermeneutic tradition' of interpretive social theory.

Framed in this way, the challenge of *re-politicizing* development entails the (re-)establishment of an interdisciplinary paradigm that is problem-oriented, action-oriented and necessarily geared towards the construction of general theory about the historical forces that affect questions of distribution, deprivation and material well-being. 'For all of its shortcomings,' Colin Leys has argued,

> ... the great merit of development theory has always consisted in being committed to the idea that we can and should try to change the world, not just contemplate it – which means, in practice, being willing to abstract from detail, to identify structures and causal relationships and to propose ways of modifying them ... 'Development studies' can no longer be conceived of as a kind of area studies, and the politics of development theory and development studies must become more explicit than ever before.
>
> (Leys 1996: 196)

However, the notion that development may reinvent itself by re-capturing history and politics understates dramatically the challenge of using theory and history to inform social action. First, the 'master disciplines' of economics, sociology and political science have moved in ways that tend to discourage 'deep' historical analysis, favouring instead a positivist emphasis on formal modeling and rational choice (Fine 2001; Flyvbjerg 2001; Harriss 2002; Kanbur 2002; Pierson 2004; Shapiro 2005; Kanbur and Schaffer 2006). Reflecting upon the state of the art of (primarily American) political science, Paul Pierson (2004) has argued that social science in the 1980s and 1990s underwent a profound 'decontextual revolution,' in which historical explanation and analysis became strongly associated with soft and woolly approaches to the study of social phenomena:

> In the midst of these strong decontextualizing trends, historically oriented analysis in the social sciences has often been criticized as a particularly egregious instance of backward thinking ... Preoccupied with contexts that are taken to be unique, historical analyses are seen as antithetical to the identification of patterns and the development of generalizations.
>
> (Pierson 2004: 168)

Although Pierson's comments were directed primarily towards political science, they could be just as easily directed towards the social sciences in general. In *Making Social Science Matter*, for instance, Bent Flyvbjerg (2001) describes a similar transformation in which social science (what Flyvbjerg calls 'episteme') has become centrally concerned with the construction of general theory inferred from hypothetic-deductive methodologies. Such methodologies, he argues, are particularly hostile to methods based on an appreciation of context and an application of value-rational theory to issues of praxis (what Flyvbjerg calls 'phronesis').

Second, 'history' is by no means an uncontested concept. As noted in the Introduction, post-colonial theorists (e.g. Spivak 1988; Kapoor 2004) question the idea that a single and objective history can exist independent of the cultural, political and epistemic biases that shape our understanding and interpretation of the past. Although he was by no means a postmodernist, E.H. Carr (1951: 9–10) shares this skepticism, and captures with great clarity the central issue at hand:

> That the Jesuit and the Marxist historian should agree about certain facts is of small importance. What matters is their agreement or disagreement on the question which facts are significant, which facts are the 'facts of history.'

History is therefore complicated by problems of interpretation on the one hand and historical consciousness and human agency on the other, highlighting fundamental questions about the ways in which history may be conducted in the context of social science research.

Such concerns raise a wider range of questions about the ways in which other histories, cultures and people may be characterized and positioned in the context of development research. Unlike 'traditional' social science disciplines, development is a field that engages with (and is often funded by) a large combination of national and international bureaucracies, whose budgets far outweigh those of conventional research funding agencies and whose commitment to basic research varies enormously from genuine interest to benign tolerance to instrumental utility (often within the same agency). Unlike many social science disciplines, development is also a field that engenders very strong feelings about poverty, suffering, inequality and injustice. Finally, and partly for this reason, there is a strong normative assumption within development that the *study* of development be intimately and essentially connected with the *practice* of development. In other words, development entails both a moral commitment to human improvement (whichever way this may be defined) and a practical commitment to achieving this aim.

However, the challenge of representing the experience of other cultures, periods and people raises a number of difficult questions about the language we use to describe the world around us and about the ways in which we engage in inter-subjective communication. As noted in the Introduction, post-colonial theory challenges the notion that the 'subaltern' may speak in relation to the academy and the West (Spivak 1988; Kapoor 2004), raising more fundamental questions about the ways in which scholars, academic disciplines, universities, research institutions and donors influence the quality and diversity of what gets studied and what gets discussed in the context of development.[6]

Arresting development? From development studies to area studies

In what follows I shall argue that development has become increasingly fragmented in terms of the theories, concepts and methodologies it uses to understand and explain complex and contextually-specific processes of economic development and social change. Outside of neo-classical economics (and related fields of rational choice), the notion that social science can or should aim to develop general and predictive theories about development has become mired in a philosophical and political orientation that questions the ability of scholars to make universal or comparative statements about the nature of history, cultural diversity and progress. The result is a field that has become extremely good at documenting the nuance and complexity of local development processes, but rather less good at connecting these ground realities to wider, historical trends and forces.

If we look at the content of what has been published in the three leading development journals (*World Development, Development and Change* and the *Journal of Development Studies*), we can see that development has been heavily pre-disposed towards the use of single country and local case study research. Between 1973 and 2008, 35 per cent of all of the articles published in *World Development* were based on a single country (see Table 1.1). For *Development and Change* the figure was 38 per cent and for *Journal of Development Studies* it was 18 per cent. Over the same period, the number of articles devoted to local or case study research increased dramatically, especially after 1990. Of the 279 'local case studies' published in *World Development*, for instance, 261 (94 per cent) were published after 1990. For *Journal of Development Studies* and *Development and Change*, the comparable figures are 64 per cent and 72 per cent (55 out of 86 for *Journal of Development Studies* and 96 out of 133 for *Development and Change*). Correspondingly, only a small number of articles published in the leading development journals (13 per cent in *World Development*, 7 per cent in *Development and Change* and 2 per cent in *Journal of Development Studies*) have used methodologies that compare two or more countries. If we extend the analysis to include 'large N' studies, the proportions drop to 5 per cent in *World Development*, 4 per cent in *Development and Change* and 5 per cent in *Journal of Development Studies*.

Table 1.1 Methodological trends in development studies: 1973–2008

Journal	World Development	Journal of Development Studies	Development and Change
Single country focus	1,369 (35%)	509 (18%)	400 (38%)
Two or more countries	528 (13%)	76 (2%)	78 (7%)
Large N	203 (5%)	132 (5%)	42 (4%)
Local/case study	279 (7%)	86 (3%)	133 (13%)
Other*	1,543 (39%)	2,006 (71%)	393 (38%)
Total	3,922	2,809	1,046

* 'Other' refers to articles that had no explicit or apparent empirical focus, and was further classified in terms of 'Theoretical/Conceptual,' 'Review Essays,' 'Methodology,' 'Commentary/Letter/Speech,' 'Policy/Organizational Analysis' and 'Industry/Infrastructure' in general.

Source: Personal survey[7]

Local research and case studies can, of course, provide important ways of connecting local realities with wider historical forces. The key point to stress, however, is that large-scale comparisons involving large bodies of data and/ or two or more countries have not featured prominently in development research (or at the very least, they have not featured prominently in the three leading development journals).[8]

The effect is 'arresting' in a number of ways. First, the notion that development may be theorized and pursued only, or primarily, on the basis of local processes appears to be too far removed from the larger structural and historical transformations that now shape the political economy of development (cf. Peet and Hartwick 1999; Fine 2001; Hoogvelt 2001; Mohan and Stokke 2000). Second, in relation to methodology and epistemology, the emphasis on local, specific processes appears to undermine the ability of scholars, researchers, activists, etc. to extrapolate insights from local and locally defined research that would, in theory, inform efforts to connect social theory and social change (Leys 1996; Mohan and Stokke 2000; Edwards 2002, 2006). Although ethnographic and case study research certainly has a role to play in the construction of knowledge (exploring, for instance, causal mechanisms, testing or investigating the validity of theories and concepts, falsifying theoretical assumptions and producing new theories and hypotheses about social processes and events), an over-riding concern is that development has become exceedingly dependent on the documentation and analysis of local and locally contingent processes and events.

To advance the debate, I will argue that development needs to recapture what Peter Evans (1995) once called the 'comparative institutional tradition.' At the heart of this approach is an inductive methodology that searches for commonalities and connections to broader historical trends and problems while at the same time incorporating divergent and potentially competing views about the nature of history, culture and development.

After Marxism: 'what is to be done?'

An important theme in this book concerns the rise and fall of the 'grand' development theories, especially ones rooted in Marxism (Leys 1996). For the better part of the twentieth century, Marxism shaped the intellectual and ideological terms on which development strategies and processes were theorized, idealized and pursued. As a body of theory, it articulated the terms and concepts that could be used to explain the transformation from primarily feudal and agrarian societies into modern and independent industrial economies. As a body of practice, it provided a model on which societies (and states leading societies) could orchestrate the economic and socio-structural changes that would lead to modernization, industrialization and development. Finally, as an ideology, it offered powerful visions of social change.

One of these ideologies was communism, whose emphasis on structural inequality and revolutionary change articulated the powerful idea that social collective action could overcome the inequality and inhumanity of colonialism, capitalism and uneven development. Reflecting upon the current state of French communism, Alain Badiou (2008) captures the essence of this powerful ideal:

> What is the communist hypothesis? In its generic sense, given its canonic *Manifesto*, 'communist' means, first, that the logic of class – the fundamental subordination of labour to the dominant class, the arrangement that has existed since Antiquity – is not inevitable; it can be overcome. The communist hypothesis is that a different collective organization is practicable, one that will eliminate the inequality of wealth and even the division of labour. The private appropriation of massive fortunes and their transmission by inheritance will disappear. The existence of a coercive state, separate from civil society, will no longer appear a necessity: a long process of reorganization based on a free association of producers will see it withering away.
>
> (Badiou 2008: 34–35)

Whether the state would lose its purpose and/or 'wither away,' the communist hypothesis had a profound effect on the kinds of nationalism and liberation being articulated and pursued during the 1940s and 1950s, reflecting the new opportunities created by an imploding imperial system.[9]

A second and related ideology was socialism, whose emphasis on public ownership and centralized planning provided a model on which states could conceivably engineer the agrarian and industrial transformations that would emulate (ideally in far less time) the industrialization of Western Europe and North America. Orchestrating these changes in the communist economies of China, Cuba and Vietnam and in the mixed economies of India, Mexico and Brazil was a bureaucratic state whose central planning apparatus could (in theory) intervene in the economy (*inter alia*, nationalizing industries, establishing quotas, setting prices, suspending private property, etc.) and allocate resources in accordance with a strategic long-term (often 5-year) plan.

Where these strategies conformed to the central aims and assumptions of Marxism, post-war development strategy therefore became strongly and necessarily associated with a large and powerful state whose ownership and control of major economic sectors could engineer the industrialization and modernization of primarily agrarian/feudal societies. Framed in this way, development was a nationalist project, whose emphasis on economic autonomy and industrial diversification legitimated the idea that Third World leaders and states would take (and plan) their societies into the modern world (Rapley 2002; McMichael, 2004; Sandbrook *et al.* 2007).

At the time, of course, the idea that states would orchestrate the management of national economic development was the mainstream of development theory and practice (Gore 2000; Rapley 2002). The Great Depression, the New Deal and the Second World War had all fostered a strong belief that the state should play a large role in economic and social life, regulating especially trade, unemployment, welfare, immigration and post-war reconstruction (Rapley 2002). Related to this, Keynesian theory (that states should stimulate employment and demand through public investment) was dominant. Moreover, the world had, by the end of the 1950s, witnessed three successful communist revolutions (in Russia, China and Cuba) whose history, ideology and politics provided a model on which other economic nationalists could pursue their own visions of development.

Today, the idea that communism, socialism or Marxism might offer a viable or desirable means of explaining, ordering or advancing society is an idea that has lost considerable ideological appeal. As Henry Bernstein has recently lamented:

> Many formerly Marxist academics, whose formation was in the 1960s and 1970s, have abandoned Marxism; there is much less Marxism available to today's university students as part of their general education in the social sciences. The connections between Marxist intellectual work and the programmes and practices of progressive political formations, both parties and regimes, have eroded with the demise or decline of the latter ... To the extent that one or another variant of Marxism exemplified a (fashionably) radical stance in the social sciences only a few decades back, this has largely been displaced by the various currents of post-structuralism, postmodernism and the like (loosely defined), the 'radical' ambitions of which rest on their subversions of the claims of existing forms of knowledge to objectivity and of any political aspirations to a project of universal emancipation.
>
> (Bernstein 2005: 126–27)

A number of factors account for this dramatic fall from grace. First, the liberalization of the command and control economies of China, Vietnam and the Soviet Union as well as the mixed economies of India, Mexico and Brazil (among many others) has undermined the perceived viability of using socialist principles of public ownership and centralized planning to pursue a state–led model of development. Second, the globalization of economic production and finance has weakened the policy instruments that central governments have traditionally used to organize and manage production and consumption, especially ones regulating tariffs, exchange rates, interest rates and the money supply, rendering ambitious *dirigiste* policies effectively obsolete (Gore 2000; Rapley 2002; McMichael 2004; Wade 2006). Third, there is now a perception that politics – and the study of politics (concerning, *inter alia*, feminism, environmentalism, religious identity, sexual identity and so on) – has moved beyond the idea that social processes may be explained only or primarily through the traditional Marxist lens (Mohan and Stokke 2000; Peet and Hartwick 1999).[10] Finally, the idea that society may be modelled and advanced on the basis of a grand social vision or ideal has been shattered (temporarily at least) by the lived experience of the twentieth century, reflecting, *inter alia*, the tragic legacy of Hitler, Stalin, Hiroshima, the Holocaust and other coordinated efforts to change the course of history (Scott 1998).

Post-structuralism, postmodernism and 'post-development'

The demise of Marxism (and modernism) is often attributed to the impact of post-structuralism and postmodernism. Although the distinctions between post-structuralism and postmodernism are by no means crystal clear (Sarup 1989; Harvey 1990; Rosenau 1992),[11] 'post-structuralism' may be broadly described as a field that explores the meaning of language and discourse to question the universalizing aims and assumptions of Western social science (including especially positivism, Marxism and structuralism). Often associated with the work of Jean-Francois Lyotard, Jacques Derrida and Michel Foucault, it implies a political and philosophical orientation that questions *inter alia* the notion that the study of language may reveal universal laws; that individuals may act with intentionality and agency; and that societies may be described in terms of binary relations, such as subject and object or, indeed, labour and capital (Sarup 1989; Rosenau 1992; Morrow and Brown 1994).

Postmodernism, on the other hand, implies a position that moves beyond the realm of social theory, and advances what is essentially an aesthetic critique of modern life and identity. Although he is by no means a postmodernist (see below), David Harvey helps to capture some of the broad contours of the field. Where 'modernism' entails the pursuit of 'linear progress, absolute truths, the rational planning of ideal social orders, and the standardization of knowledge and production' (Harvey 1990: 7), Harvey argues, postmodernism 'privileges heterogeneity and difference … (f)ragmentation, indeterminacy, and intense distrust of all universal or "totalizing" discourses' (1990: 9).

Framed in this way, the 'postmodern condition' is one in which narratives of salvation, liberation, enlightenment and development are treated with deep suspicion and incredulity, and any effort to define 'art, morality and science' is necessarily contingent upon one's own interpretation of what is beautiful, what is good and what is true (Sarup 1989: 132). In *The Postmodern Condition*, Jean Francois Lyotard (1984) suggests that modernism embodies a 'teleology of progress' whose underlying aim (or meta-theory) is defined primarily in relation to the emancipation of groups and individuals repressed by the weight of ignorance, religion, capitalism, history, etc. Although liberating, Lyotard suggests, the aims and objects of emancipation are unavoidably subjective, and produce a 'totalizing discourse' that emphasizes and excludes the liberation of certain needs and worldviews over others (Parfitt 2002; Sarup 1989). For Lyotard, modernism entails:

> ... any science that legitimates itself with reference to a metadiscourse ... making an explicit appeal to some grand narrative, such as the dialectics of the spirit, the hermeneutics of meaning, the emancipation of the rational or working subject or the creation of wealth.
>
> (cited in Sarup 1989: 131)

Postmodernism therefore extends the post-structural critique of knowledge and language to include the aims, achievements and aesthetics of modernity. It is '*post*-modern' in the sense that it questioned fundamentally the idea that rational planning and coordination may yield universally desirable forms of human progress.

Influenced by postmodernism, post-structuralism and (especially) Foucault, 'post-development' challenges the idea that development may be understood as a benevolent response to the manifest needs of the world's poor, suggesting instead that 'development' is an idea whose primary function is to serve the commercial and geo-political needs and interests of capitalism, globalization and 'the West.' Framed in this way, development constitutes a domain of power, whose definition, measurement and treatment of development 'problems' (like poverty, hunger, malnutrition, pain and suffering) mask and legitimate what is in fact the application of American/Western/bureaucratic power.

For scholars wedded to the ideals and approaches of Marxism, the postmodern challenge to meta-theories of knowledge, reason and progress fosters a relativism that appears to trivialize the universalizing tendencies of capitalism, undermining the ability of social classes (and of intellectuals) to effect social change.[12] 'In pulling the rug out from under the certainties of its political opponents,' Terry Eagleton (1997: 24) has argued, 'this postmodern culture has often enough pulled it out from under itself too, leaving itself with no more reason why we should resist fascism than the feebly pragmatic idea that fascism is not the way we do things in Sussex or Sacramento.' Similarly, Ellen Meiksins Wood (1997: 13) has worried that

> ... the postmodernist insistence that reality is fragmentary and therefore accessible only to fragmentary 'knowledges' is especially perverse and disabling. The social reality of capitalism is 'totalizing' ... and an understanding of this 'totalizing' system requires just the kind of 'totalizing knowledge' that Marxism offers and postmodernists reject.

Similarly, from a liberal perspective, Charles Taylor (1991; 2007) has argued that postmodernism (and here he is referring specifically to Derrida and Foucault) understates the ability of individuals to orient one's own life and one's life experience to something that goes beyond the self. 'What is peculiar to the modern world,' he suggests,

... is the rise of an outlook where the single reality giving meaning to the repeatable cycles (of time, life, tradition, routine, etc.) is a narrative of human self-realization, variously understood as the story of Progress, or Reason and Freedom, or Civilization or Decency or Human Rights; or as the coming to maturity of a nation or culture ... The meaning of these routines, what makes them really worth while, lies in this bigger picture, which extends across space, but also across time.

(Taylor 2007: 716)

Taken to the extreme, Pauline Marie Rosenau (1992) has argued, 'skeptical postmodernism' (and here she is including the work of Derrida, Lyotard and Foucault) challenges the idea that history and identifiable historical periods can ever be used to explain social phenomena. First, any effort to periodize the past is bound to reflect the subjectivities and biases of dominant ideological and intellectual forces. Such biases may be the result of conscious efforts to construct a particular version of the past. Or they may simply be the result of memories, which are imperfect and subject to one's own partial recollection of the past. Second, efforts to compare different time periods assume that the conditions that gave rise to an event occurring at one period in time are the same as (or as comparably influential as) ones occurring at another.

More radical positions would question the idea that systemic patterns or forces cut across different periods in time, and suggest instead that every experience or 'moment' is, by its nature, indefinable and unique (Rosenau 1992: xii). Framed in this way, the existence of an epoch or *longue duree* suggests an existence that transcends and explains the ideas, actions and institutions of a particular point in time (Jameson 1984; Sarup 1989; Harvey 1990; Rosenau 1992). By the same token, and at the other extreme, the notion that singular individuals or events may represent or account for critical points or junctures overstates the agency of individual personalities and events (Sylvester 1999). The result, Frederic Jameson (1984) has argued, is a 'schizophrenic condition,' in which efforts to understand or change the course of history are usurped by a profound sense of rupture and fragmentation.

Whether the authors that Rosenau (1992) associates with 'postmodernism' can be characterized in this way, such positioning creates significant problems for a social science that aims to construct, through history, universal insights about economic development and social change. Primarily it denies the ability of history to exist in isolation from the values, biases and ideals of historical interpretation. Secondarily, it challenges the idea that inferences may be established on the basis of past events, undermining the idea that lessons can be learned – and future behaviour modified – on the basis of historical processes and events.

Therefore, an overriding concern is that postmodernism has narrowed the ability of individuals, ideas and ideologies to theorize and advance the human condition. A related concern is that postmodern renderings of knowledge, power and progress have narrowed the terms on which alternatives to neo-classical theory and neo-liberal development may be theorized, idealized and pursued. Tom Brass, for instance, has argued that postmodernism 'transformed the disillusion of 1968 into a Nietzschean pessimism, which licenses and, in politico-ideological terms, epitomises the reactionary conservatism of the 1980s' (Brass 1991: 177). Whether the 'postmodern' turn facilitated the 'neo-classical' turn, Brass certainly helps to capture a wider concern that postmodernism and neo-liberalism have fostered what Frederic Cooper and Randall Packard have called a 'strange convergence of free market universalists and anti-universal critics' (Cooper and Packard 1997: 4).

The notion that development normalizes, 'de-politicizes' and excludes (even when it assists) is a powerful one, and one that has captured a literature that goes well beyond postmodernism (e.g. Harriss 2001). However, taken to the extreme, the notion that development is inherently biased against the poor, against the 'South,' etc. fosters a nihilism that questions not only the dominant discourses and practices of the IMF, the World Bank, etc., but also calls into question any effort to improve the human condition. In so doing, it fails to articulate with sufficient clarity or purpose the ways in which it would move beyond the dominant and/or neo-classical development paradigm. As Stuart Corbridge has argued,

> ... post-developmentalism puts us in touch with the victims of development (and the authors of alternatives to development) in a way that escapes the under-socialised accounts of human action that find favour in development economics. And this matters. It would be absurd to reduce development studies to a sort of generalized moral indignation (and such an absurdity is present in post-development thinking), but it is no less absurd to reduce development issues to a positive social science which obsesses about means and only rarely considers the ends of 'development.'
>
> (Corbridge 1998: 143)

Therefore, a critical understanding of the politics and ideology of development entails an open inquiry about the nuance and diversity of what constitutes the 'dominant discourse.' Towards this end, this book aims to advance a critical dialogue about the ways in which neo-liberalism and related worldviews of neo-classical theory and rational choice have framed and defined the 'meta-theoretical' aims and assumptions of what is deemed relevant, important and appropriate to the study of development. In so doing, it aims to explore the theoretical and ideological terms on which alternatives to neo-classical theory may be theorized, idealized and pursued.

Outline of the book

So how does the book proceed? Chapter Two first situates historical and neo-classical research in a wider context of positivist social science. Drawing upon new institutional theories of collective action and rational choice, it explores the meta-theoretical assumptions on which neo-classical and historical approaches are established, validated and defended in the context of social science research. In so doing, it makes the case that the meta-theoretical differences that divide historical and new institutional approaches to the study of poverty, livelihoods and the commons are essentially differences between a social science that seeks to build theory on the basis of scientific empiricism and an ethnography that rejects the universalism that underlies the scientific approach.

Chapter Three explores the historical and theoretical factors that, during the latter half of the twentieth century, began to question the (once powerful) idea that history may conform to a set of theoretical assumptions about the dialectical nature of capitalism, class struggle and revolutionary change, undermining for many the notion that social classes (*and intellectuals*) may effect progressive social change. Underlying the retreat from the 'grand' theoretical narratives of Marxism and dependency was both a rejection of the more deterministic efforts to establish a science of history and a more fundamental desire on the part of scholars like James Scott (1985) and Eric Wolf (1982 [1997]) to give the objects of social science (i.e. people) the opportunity to define through their own interpretations and practice the terms on which their values, experiences and histories were used for the purposes of social science scholarship.

Chapter Four explores the impact of postmodernism (or Foucault at any rate) on the field of 'post-development.' At the heart of this philosophical and political orientation is the notion that development may be discerned through the 'discourses' that structure and give meaning to development policy and practice. In so doing, it aims to expose the unstated aims and assumptions of development, which include both the old development orthodoxies of modernization, as well as the new development orthodoxies of neo-classical economics and neo-liberal reform, thereby illuminating the historical and epistemic conditions under which 'development' emerged as a body of theory and practice.

Raising concerns about the viability of using Marxism and/or postmodernism to advance a clear and coherent vision of change, Chapter Five then considers the seemingly counter-intuitive idea that liberalism may provide an ideological and theoretical alternative to the neo-classical frame. Perhaps the most ambitious argument in favour of a liberal democratic discourse can be found in the work of Amartya Sen, whose 'capabilities approach' articulates the powerful idea that development entails more than improvements in national or personal income, but also the freedom to choose the life one wants to pursue, and to establish, through politics and open discourse, the moral terms on which life, development and therefore freedom may be understood and assessed. Chapter Five considers the ways in which and the extent to which Sen's conceptualization and related efforts to introduce participatory principles in development research may capture an image of development that is less beholden to the formal assumptions of neo-classical theory.

Finally, Chapter Six concludes the book by situating these trends in a wider historical context and outlining a number of ways in which comparative social theory may advance a theoretical and ideological alternative to the neo-classical frame.

2 The 'poverty of history' in neo-classical discourse

Positivism, new institutionalism and 'the tragedy of the commons'

I did not possess you, but I can blow up history.
                    Umberto Eco, *Foucault's Pendulum* (1989: 435)

History is more or less bunk.
                                                        Henry Ford

Introduction

The previous chapter introduced the idea that neo-classical theory has influenced the meta-theoretical aims and assumptions of development theory and research. This chapter now explores the impact of neo-classical theory on a more specific literature concerning the management of natural resources, collective action and 'the tragedy of the commons.' Our central aim is to consider first the tensions that exist between new institutional and historical approaches to the study of poverty, livelihoods and the commons, and second, the arguments marshalled for and against the practice of generalization in the social sciences.

As noted in Chapter One, the new institutionalism constitutes an important response to neo-classical theory. Assuming that individuals are 'bounded' in their ability to find, assess and acquire the goods and services that would satisfy their (relatively stable) preferences, it marks an important departure from the idea that markets will emerge naturally to coordinate 'with varying degrees of efficiency' (Gilpin 2001: 51) the needs and preferences of utility maximizing individuals. However, as Geoffrey Hodgson (1993) has pointed out, the new institutionalism is still firmly wedded to the assumption that social outcomes may be understood and explained primarily on the basis of individual decision making and rational choice. Indeed, Hodgson (1993) has argued that a commitment to the rational choice theoretic is a defining factor that differentiates the 'new institutionalism' of Douglass North (1990) from the 'old' historical institutionalism of Thorsten Veblen (Hodgson 1993).

In what follows, I shall argue that 'new institutional' approaches to the study of the commons share with neo-classical theory a desire to understand and model society on the basis of individual decision making and rational choice. At the heart of this mainstream is a wider trend of positivism, methodological individualism and formal modelling that has come to dominate American political science (a discipline that is home to leading figures in the field, including Elinor Ostrom, Arun Agrawal and Robert Wade). The 'poverty of history' to which I refer in the title highlights two themes that underlie this chapter. The first theme refers to the gap that exists between a literature that uses the past to develop general and predictive theories about what makes for durable common property regimes and one for whom the central questions and mechanisms of change are far more contextually specific.[13] The second theme represents a wordplay on Popper's (1957 [1997]) classic assault on 'historicist' thinking within the social sciences. Drawing upon the ideas developed in this thesis, I aim to explain the appeal of the scientific method on which the new institutionalists now make their most assertive claims, and the constraints that therefore prevent them from adopting a more historical mode of analysis.

The chapter proceeds thus: the following section first outlines the methodological and meta-theoretical differences that exist between historical and positivist social science. Next I situate Hardin (1968 [2005]) and his critics in relation to neo-classical scholarship, arguing that a preoccupation with efficiency and the management of tragedy of the commons scenarios constitutes the mainstream of new institutional work on environmental conflict, collective action and the commons. Third, I explore the tension that exists between scientific and historical explanation, and make the case that 'scientific approaches' to the study of the commons are strongly influenced by – and deeply embedded in – a positivist social science, which has become particularly influential in American political science and the 'new institutional economics.' Finally, at the conclusion of the chapter, some general observations are made about the implications such tensions pose for the study of development.

The 'problem' of history in social science research

Debates about theory and methodology in the social sciences have long been concerned about the extent to which scientific principles may be applied to the study of social phenomena. The 'problem' of using history to understand the social world is that social phenomena do not lend themselves very well (or at all, some would say) to the central aims and assumptions of science (Collingwood 1946 [1992]; Carr 1951; Moore 1966; Hobsbawm 1987 [1989]). For one, the notion that historical processes and events may be 'isolated' and tested as if they were physical properties and processes understates substantially the possibility that the factors that led to one sequence of events and outcomes were completely unique, and that they can therefore never be compared or replicated again. In the words of one American scholar, historical narratives are like 'Seussian explanations,' in which 'it just happened that this happened first, then this, then that, and is not likely to happen that way again' (Jack Goldstone, cited in Pierson 2004: 169).

A second issue concerns the extent to which facts may be treated in isolation from the norms, assumptions and values we (i.e. researchers, respondents, students, readers, etc.) use to give them meaning (Morrow and Brown 1994; Flyvbjerg 2001;

Putnam 2002). As Bent Flyvbjerg (2001) has argued, attempts to emulate the natural sciences in the social sciences are problematic in the sense that scientific inferences about human behaviour can never approximate the 'context-dependent' factors that determine human motivation:

> The problem in the study of human activity is that every attempt at a context-free definition of an action, that is, a definition based on abstract rules or laws, will not necessarily accord with the pragmatic way an action is defined by actors in a concrete social situation.
>
> (Flyvbjerg 2001: 42)

Third, as noted in Chapter One, the events of history are also subject to interpretation and manipulation. As E. H. Carr has argued, the idea that human societies may be studied 'as if' they were bound by the laws of physics underplays the notion that the subjects of history are conscious of their surroundings and of their past, and that their consciousness may affect the course of history. In his classic essay on *The Historical Approach*, he describes the problem as follows:

> In science the drama repeats itself over and over again because the *dramatis personae* are creatures unconscious of the past or inanimate objects. In history, the drama cannot repeat itself because the *dramatis personae* are already conscious of the prospective *denouement*; the essential condition of the first performance can never be reconstituted.
>
> (Carr 1951: 5–6)

For Carr (1951), a central issue of concern was the Enlightenment's faith in the idea that society and history could be explained (and ordered) through the application of science and reason. Where natural scientists collected observations about physical processes and properties, he argued, the 'modern historian' aimed to establish through the 'events of history' basic and universal insights about the laws of progress, evolution and change. Influenced by the liberal norms of the French Revolution, for instance, Montesquieu and Condorcet looked into history to find the causal forces that would lead to human liberation and progress (Carr 1951). After the publication of *The Origin of the Species*, historians began to look for processes and events that would conform to Darwinian assumptions about natural selection and evolutionary change. Finally, in the wake of the Holocaust and the Second World War, the historical context in which Carr was writing, the historical leitmotif became one of destruction, civilization and the end of empire (Carr 1951).

The basic problem of using scientific and evolutionary analogies to understand history, Carr (1951) argues, is first that historical trends and forces cannot be replicated, and are therefore always subject to the values, subjectivities and classifications used to interpret the past. Second, although human relations certainly embody elements of power and intention, the notion that history may move or unfold in one way or another raises questions about the ways in which and the degree to which these forces may be influenced by conscious efforts to avoid, replicate or repeat the past (also see Chapter Three).

History and positivist social science

Although scholars have, in recent years, tried to incorporate history into a scientific frame (see Chapter Six), the dominant strategy within positivism has been to resolve problems of interpretation and consciousness by relegating history to the margins of social analysis (Morrow and Brown 1994; Pierson, 2004). In Morrow's words,

> Knowledge about history is held to have no significance for the evaluation of the validity of theories and to be largely peripheral for the discovery of better research strategies.
>
> (Morrow and Brown, 1994: 67)

Perhaps the most influential articulation of this kind comes from Karl Popper, whose *Poverty of Historicism* (1957 [1997]) shared with *some* positivists the view that historical narratives are exceedingly dependent upon personal biases and subjectivities,[14] and that they preclude the testing and falsification of factors other than the ones highlighted by the individual historian:

> (The historicist) firmly believes in his favourite trend, and conditions under which it would disappear are to him unthinkable. The poverty of historicism, we might say, is a poverty of imagination.
>
> (Popper 1957 [1997]:129–130)

Underlying this approach is a foundational belief that historical narrative can be incorporated into a scientific frame, in which the inherent bias of historicist thinking is eliminated while still maintaining the context of past events.

Popper's case against 'historicist' explanation stems from the distinction he makes between historical accounts that look for regularities and 'innumerate facts … in some kind of causal fashion,' and those that argue that 'unique events … may be the cause of other events' (Popper 1957 [1997]: 146). In a thinly veiled assault on Marx's and *Marxist* historiography, Popper argues that historicist accounts, in which causal explanations are made on the basis of 'unseen' laws of motion, are inconsistent with a scientific (and therefore falsifiable) understanding of reality:

> The attempt to follow causal chains into the remote past would not help in the least, for every concrete effect with which we might start has a great number of different partial causes; that is to say, initial conditions are very complex, and most of them have little interest for us.
>
> (Popper 1957 [1997]: 150)

How one forms an original question or hypothesis was a matter of some difficulty for Popper (cf. Hollis 1994), rectified only partially by his belief that human beings are 'born with expectations,' which are 'prior to all observational experience.' The most important of these, he argued, was the expectation of 'finding a regularity' (cited in Hollis 1994: 74). In his own words,

> I do not believe that we ever make inductive generalizations in the sense that we start with observations and try to derive our theories from them. I believe that the prejudice that we proceed in this way is a kind of optical illusion, and that at no stage of scientific development do we begin without something in the nature of a theory, such as a hypothesis, or a prejudice, or a problem – often a technological one – which in some way *guides* our observations, and helps us to select from the innumerable objects of observation those which may be of interest.
>
> (Popper 1957 [1997]: 134)

Where experience and observation contradict our understanding of what constitutes normal or regular behaviour, Popper argues, we have *a problem*, for which new explanations are required:

> We try; that is, we do not merely register an observation, but make active attempts to solve some more or less practical and definite problems. And we make progress if, and only if, we are prepared to *learn from our mistakes*: to recognize our errors and to utilize them critically instead of persevering in them dogmatically.
>
> (Popper 1957 [1997]: 87; italics in original)

Leaving for now questions about who defines what constitutes 'a problem,' Popper draws our attention to the idea that a hypothetic–deductive model may be used to frame and interpret history (and other social phenomena).

Broadly speaking, positivism tends to eschew the idea that social realities may be established on the basis of metaphysical causal forces, embracing instead the notion that the only (or primary) means of understanding the world is to identify patterns and regularities in empirical data. As Martin Hollis has argued,

> The *methodology* is aimed at identifying regularities in the behaviour of particulars. It does not seek to detect underlying structures, forces or causal necessities, for the good reason that there are none. It involves theoretical abstraction and deductive reasoning but only for the sake of arriving at improved predictions. Inductive generalisations do the crucial work ... The *epistemology* is as basic and simple a version of empiricism as will warrant the governing precept that only perception and the testing of prediction can justify claims to knowledge of the world.
>
> (Hollis 1994: 64)

Starting from the proposition that metaphysical assumptions about the external world are inherently unreliable, positivism therefore aims to structure the acquisition of knowledge in a way that emulates the basic principles of science. At the heart

Empiricism is a related theory of knowledge that emphasizes experience and sensory perception as a principal means of explaining social phenomena. An axiomatic assumption is that the words, theories and concepts we use to describe the world may exist in isolation from the facts that constitute reality, suggesting a distinction between what David Hume once called 'relations of ideas' and 'statements of matters of fact,' (cited in Putnam 2002: 13). Framed in this way, knowledge may be established on the basis of 'sense data,' essentially 'the things that are immediately known' through basic human sensation (Kanbur and Shaffer 2006: 186).

Box 2.1   Empiricism

of the scientific method is the notion that explanations about social and natural phenomena may be established on the basis of a methodology that adheres to principles of validity and reliability. Questions about validity concern the ability of concepts and methods to measure what they intend to measure (Peters 1998). Reliability is the principle that 'applying the same procedure in the same way will always produce the same measure,' even when it is applied by different researchers (King, Keohane and Verba, 1994: 25).[15]

Whether inferences are made on the basis of deductive or inductive reasoning (or a combination of the two), the basic requirement is that the methodology adhere to a common set of principles (of deduction or induction) which can then be tested and replicated in multiple research settings.

In short, the methodology being advanced by positivism is one in which the assumptions, propositions and conclusions of formal theories are tested both in terms of their logical coherence and their consistency with empirical data. Framed in this way, knowledge is rooted in the practice of empirical observation, the falsification of testable hypotheses, the inconclusive nature of theoretical statements, the central importance of method and the 'public' nature of social science and the 'scientific community.' Therefore, the basic aim of social science research is to structure social inquiry in a way that is oriented towards the generalization of insights about social phenomena (cf. Johnson 2006).

Debating positivism

Debates about the viability and desirability of using positivism to structure the acquisition of knowledge first concern the assumptions we make about human behaviour and second, the ways in which we convert particular observations into general insights. Following Popper (1962), the notion that hypotheses may be tested and *confirmed* in social science research settings understates dramatically the challenge of isolating and evaluating the impact of multiple causal variables. Popper's (1962) argument was that any theory will generate hypotheses that are so vast in number that they preclude testing and verification. His solution was therefore to advance the notion that hypotheses can only be 'falsified' in relation to existing knowledge or 'conjecture' about the social world. Similarly, Thomas Kuhn (1962) argued that findings that contradict the predictions and regularities described by existing theories are said to contribute to knowledge by disproving or falsifying the assumptions of what we hold to be true.

The value of using positivism to structure the acquisition of knowledge is that it forces us to lay bare the assumptions we make about social phenomena, and subject them to critical scrutiny (cf. Laitin 2003). The drawback, of course, is that it pre-determines the questions, variables and hypotheses in question, creating an artificial distinction between the subjects and objects of social science research. First, the assumption that positivist methodologies may be developed in isolation from the object of social science inquiry is clearly at odds with the view that language, concepts and 'covering laws' provide only a partial understanding of individual subjectivities (Morrow and Brown 1994; Flyvbjerg 2001; Johnson 2006). At the heart of this critique is a deep scepticism about the norms and assumptions on which researchers and respondents engage in positivist social science research. As John Harriss has argued:

> The sorts of answers that people give to survey questions about attitudes and values may be interesting but there is very often a lot of doubt as to how 'respondents' have understood the questions which are posed to them, and how their answers are influenced by the context in which they are interviewed.
>
> (Harriss 2002: 489)

Whether the respondents understand or disagree with the questions being asked of them seems largely irrelevant. Moreover, the notion that reliability can be defined and achieved by replicating the methods and measurements laid out in positivist research methodology substantially underplays the subjectivities (of style, experience, *language*) that individual researchers may bring to the research process.

Second, there is a problem of context. As Paul Pierson has argued, the problem of using this kind of social inquiry is the notion that it simplifies, misrepresents and excludes the complexity, diversity and details of history:

> Too often, contemporary social science simply drops out a huge range of crucial factors and processes, either because our theories and methods make it difficult to incorporate them, or because they simply lead us not to see them in the first place.
>
> (Pierson 2004: 169)

Third, the idea of moving to the general from the particular gives the impression that inference and generalization are the *only* meta-theoretical goals of social inquiry. However, as James Johnson (2006) has argued, the assertions being advanced by positivism 'obscure' the nature and utility of qualitative research, reducing to quantifiable points of data the perceptions and insights of social science respondents, whose cumulative purpose is to test, support or refute research hypotheses (cf. George and Bennett 2004; Flyvbjerg 2001; Harriss 2002).[16]

Debates about positivism therefore raise a number of fundamental questions about the meta-theoretical aims of history and of social science research. First, do we study history in order to understand a particular event or series of events? Or do we aim to generate conclusions that can be used to understand a wider class of phenomena? Second, do we develop our understanding of these phenomena on the basis of inductive reasoning whereby observation and investigation are followed by classification and theorizing? Or do we first classify and theorize causal relations a priori and *then* investigate them in order to confirm or falsify their validity? The first question is essentially about the ethics and aesthetics of primary research. Do we investigate the Russian Revolution because we want to know how or why the Russian Revolution happened, or do we study it because we want to know why revolutions (in general) happen? The second concerns the means by which we arrive at our impressions about the world around us.

The following section explores the meta-theoretical tensions that exist between new institutional and historicist approaches to the study of the commons, making the case that 'new institutional' approaches to the study of the commons have been strongly influenced by – and deeply embedded in – neo-classical theory and positivist social science, which has become particularly influential in American political science.

The tragedy of the commons

Writing in 1968, Garrett Hardin published what would become an influential and highly controversial essay about the causes and nature of environmental problems, in which he argued that environmental pollution and resource depletion are primarily the result of three fundamental 'gaps' that exist between the utility individuals are assumed to derive from extracting and polluting natural resources, and the challenge of ensuring that resources are collectively managed and maintained. One relates to the distribution of social cost. Reflecting upon the challenge of managing an open field system of grazing, Hardin's argument (1968 [2005]) was that resources owned by no one (*res nullis*) would by necessity fall prey to over-extraction because they allow individuals to extract resources without bearing the costs of managing and maintaining the resource. Lacking rules of management or private property individuals will therefore overuse or pollute the system when the (personal) benefits of overgrazing or pollution outweigh the (personal) costs of exercising restraint. Therein, Hardin asserts, lies the tragedy.

A second gap relates to the information that individuals have about the state of the resource. An important concern here is that individuals, lacking reliable information about the quality and availability of renewable resources, will continue pumping or depleting them until they are beyond repair. Take, for instance, the case of a fishery. The challenge of managing a fishery stems from the fact that a fishery can only be maintained if there is a mature standing stock, which is left to reproduce after seasonal or perpetual harvests. Although particular species will breed, spawn and mature in certain areas, their position within these areas can vary greatly and exogenous shocks (such as changes in rainfall, variable sunlight, water pollution) can disrupt these patterns considerably. Both of these factors foster uncertainty in the sense that they make it very difficult to predict how many fish (and therefore, how much income) one will bring in on any given day. This, in turn, creates strong incentives to pull in as many fish as possible today because there may be none tomorrow or (significantly) because your competitors may beat you to it. It also creates incentives to invest in labour and technology that will reduce the uncertainty and risk that your boat will come home with empty nets. A final, related form of uncertainty stems from the information that individuals have about those with whom they share resources.[17] Lacking credible information about the intentions and activities of other resource users, Hardin (1968 [2005]) argued, individuals will inevitably exploit the commons, fearing other self-seeking individuals will inevitably do the same.

Starting from the proposition that environmental degradation is the result of individuals having unrestricted access to 'open access resources,' Hardin (1968 [2005]) argued that the only viable solution to the tragedy of the commons was to restrict the terms upon which people use and obtain access to the commons. One way of doing this, he maintains, was to create rights of private property which would, in theory, give individuals the incentive to manage and maintain resources from which they could conceivably reap future benefits. A second and more radical option was to restrict individual freedom, including 'the freedom to breed.' (Hardin 1968 [2005: 31])

Avoiding the tragedy: institutions, incentives and 'common property regimes'

In critical response to Hardin (1968 [2005]), new institutional approaches to the study of the commons have documented, tested and explained literally thousands of cases in which people have been able to organize what Schlager and Ostrom (1992) usefully classified as rules of access, withdrawal, management, exclusion and alienation, thereby creating and/or maintaining common property regimes. At the heart of this 'renaissance' is the assertion that variations in property regimes and relations have strong bearing on the ways in which people use, manage and abuse natural resource systems, and that institutional arrangements – grounded in the creation and management of *common* property – can have positive impacts on resource use and conservation.[18]

The logic of the argument is that individuals will be more likely to conserve a resource when they believe they will reap the long–term benefits of conservation and restraint. Common property, it is argued, provides this assurance by restricting otherwise open-access resources to a group that agrees to abide by rules regulating membership and resource utilization (Baland and Platteau 1996; Bromley *et al.* 1992; Ostrom 1990; Uphoff *et al.* 1990; Wade 1988). A critical distinction was the notion that 'common property regimes' (CPRs) had rules regulating the ways in which individuals obtain access to natural resources, thereby eliminating the open access dilemma (Bromley *et al.* 1992; Libecap 1995; Ostrom 1990; Wade 1988).

As McCay and Jentoft (1998) have argued, 'commons scholarship' underwent an important transformation after Hardin articulated his original thesis in 1968. Prior to Hardin, 'commons dilemmas' were understood either in terms of Malthusian explanations of overpopulation and resource degradation (McCay and Jentoft 1998) or in terms of an agrarian transition to capitalism rooted in the enclosure of the English commons and the creation of an incipient working class (e.g. Marx 1867 [1976]; Thompson, 1963: pp. 237–58).[19] In McCay and Jentoft's words, the central focus of this new scholarship shifted from one of poverty and overpopulation to one of efficiency and environmental conservation:

> The model that Hardin revived was taken up by students of institutional and natural resource economics and the evolution of property rights. The question changed from why so many poor people to why natural and economic resources were wasted and depleted.
>
> (McCay and Jentoft 1998: 21)

Such insights were largely dependent on a seminal literature (e.g. Acheson 1988; Dahlman 1980; Netting 1981; McCay and Acheson 1987; Ostrom 1990), which emerged in the early 1980s, and established that many of the commons Hardin (1968 [2005]) had understood to be open access were in fact governed by rules of common property. Drawing upon historical material and their own field research, scholars like Elinor Ostrom (1990) and Robert Wade (1988) argued that Hardin had underplayed the possibility that what he was calling open access could in fact be governed by common property.

Whether common pool resources would in fact fall prey to the tragedy of the commons was therefore not a matter of faith as Hardin (1968 [2005]) would lead us to believe, but an empirical question, dependent on the existence of institutions governing access, utilization, management, exclusion, ownership and transfer of ownership (Schlager and Ostrom 1992), and on the ways in which individuals respond to these structural incentives.

Because they employed a unified set of concepts and methods that could *in theory* be applied in different geographical and socio-economic settings, rational choice theories (RCTs) offered an attractive means by which researchers might infer more general insights about the conditions under which individuals and institutions would create the incentives to establish and conserve common property regimes (e.g. Wade 1988; Ostrom 1990; Baland and Platteau 1996). For Elinor Ostrom, research could inform policy by comparing successful and unsuccessful efforts to establish common property regimes:

> Instead of basing policy on the presumption that the individuals involved are helpless, I wish to learn more from the experience of individuals in field

settings. Why have some efforts to solve commons problems failed, while others have succeeded? What can we learn from experience that will help stimulate the development and use of a better theory of collective action – one that will identify the key variables that can enhance or detract from the capabilities of individuals to solve problems?

(Ostrom 1990: 14)

Starting from the position that information (about resources and resource users) is costly, the new institutionalists contend that rules *matter* because they reduce the uncertainty that stems from the unpredictable behaviour of 'resource users' and resource systems. As Ostrom (1990) puts it,

> In all cases in which individuals have organized themselves to solve CPR problems, rules have been established by the appropriators that have severely constrained the authorized actions available to them. Such rules specify, for example, how many resource units an individual can appropriate, when, where, and how they can be appropriated, and the amounts of labor, materials, or money that must be contributed to various provisioning activities. *If everyone, or almost everyone, follows these rules, resource units will be allocated more predictably and efficiently, conflict levels will be reduced, and the resource system itself will be maintained over time.*

(Ostrom 1990: 43; my emphasis)

Collective action scholarship on the commons is therefore principally interested in the ways in which individual incentives and social institutions combine to affect social outcomes. Such outcomes are explained both in terms of individual calculations and structured incentives. An important point made by scholars like Ostrom (1990), Wade (1988) and Baland and Platteau (1996) is that structure, group attributes, resource attributes and ecological pressures affect the ways in which individuals understand and respond to different social dilemmas (see especially Agrawal 2001; and below). However, at the core of the collective action school is a model of individual decision making and rational choice.

Influenced by new institutional theories of collective action and rational choice, scholarship on the commons embraced a new meta-theoretical orientation, which encouraged an empirical investigation of the conditions under which individuals could cooperate to govern the commons. Three assumptions were central to this model. One was the notion that social outcomes can be explained in terms of the calculations that individuals make about the perceived costs and benefits of future actions (methodological individualism). A second was that individuals are 'rule-governed' in this process (see below). A third and critical component was a substantive re-conceptualization of the commons, which had the important effect of challenging Hardin's thesis and enabling a new empiricism in common property research. An important distinction was made between common pool resources and common property regimes (Ostrom *et al.* 1994; Ostrom 2000). The former referred to resource systems whose size, mobility and complexity make it difficult – although not impossible – to prevent individuals from using them and whose use can deplete the number and quality of benefits the resource can provide. The latter were the systems of rules, rights and duties that govern the ways in which group members relate to the commons and to one another (Ostrom 2000).[20]

The new scholarship also entailed a new methodological agenda, which had its roots in the rejection of behaviouralism (Immergut 1998), and the development of a hypothetic–deductive model rooted in neo-classical assumptions about individual decision making and rational choice. Drawing upon theories of collective action and public goods (e.g. Olson 1965), this agenda was centrally concerned with the problem of understanding and *developing principles which would encourage* collective action to conserve local resources. Framed in this way, environmental problems were understood as collective dilemmas, in which individuals deplete resources because they lack: (1) information about the resource system; (2) information about those with whom they share the resource; and (3) rules that would regulate the ways in which they use the resource.

In short, rational choice theories of collective action to conserve the commons developed what would become a highly influential approach to the study of individuals, incentives and natural resource management. However, the literature that began to crystallize on the connections among property rights, property relations and the commons was not without dispute. In what follows I shall argue that two bodies of thought compete for a voice in this literature. One, responding primarily to Hardin's tragedy of the commons (1968), is concerned with the problem of achieving collective action to regulate and conserve natural resources that are prone to depletion. A second, influenced by notions of moral economy (e.g. Thompson 1971; Scott 1976) and entitlement (Sen 1981), deals with the problem of creating and sustaining resource access for poor and vulnerable groups in society (Beck 1994; Jodha 2001; cf. Mosse 1997).[21]

The two schools converge in the sense that they both use rules or 'institutions' to analyze and describe the dilemma of managing and obtaining access to natural resource systems. However, they differ in the normative value that they ascribe to common property regimes and those whose livelihoods are dependent on the resources it provides. Whereas collective action scholars analyze the rules and sanctions that encourage individuals to conserve the commons, 'entitlement scholars' emphasize the historical struggles that determine resource access and entitlement, and the ways in which formal and informal rules create and reinforce unequal access to the commons. There is thus a normative tension within the literature.

There is also a methodological tension. Although the lines are by no means neatly drawn, entitlement scholarship on the commons tends to favour an interpretive–historical approach, in which transformations of property rights regimes are explained principally in terms of historical narrative and context. In contrast, collective action scholars (such as Ostrom 1990; Wade 1988; Baland and Platteau 1996) embrace a deductive model of individual decision making and rational choice to explain the ways in which different types of property rights arrangements emerge and change over time (e.g. Ostrom 1990) and space (e.g. Wade 1988). To the extent that the past is used to inform these inferences, the collective action school adopts a historical method. However, the historiography that underlies this approach adopts a view of the past that subsumes the peculiarities of history to the construction of general theories and the evaluation of these theories in the field. As Paul Pierson has argued, the process is one in which historical 'facts' are selected and interpreted to inform theory on which future behaviour can be based:

> Most rational choice analysts place priority on combining the greatest degree of parsimony and the greatest capacity for generalization. This leads to a presumption that compelling hypotheses involve little in the way of 'local' information.
>
> (Pierson 2004: 168)

In contrast, the historiography and historical methods that underlie the entitlement school are centrally concerned with past and contextually specific questions, changes and events. To quote Geertz (1973: 5), the entitlement school is 'not an experimental science in search of law, but an interpretive one in search of meaning.' In defence of their method, entitlement scholars have criticized the collective action literature for its instrumental and historically de-contextualized understanding of common property relations, calling for a more historical understanding of the ecological and socio-economic factors that affect the myriad relations on which property, common property and other forms of resource entitlement are based (see especially, Campbell *et al.* 2001; Cleaver 2000; Goldman 1998; Mosse 1997; Prakash 1998). Others such as Scoones (1999) have argued for a deeper understanding of the complexity, uncertainty and dynamics that underlie ecological processes and environmental change.

The poverty of history

Although rational choice scholars use historical material to construct their assertions, their use of history – and the historiography that underlies their epistemology – differs from the historical approach that tends to be favoured among entitlement scholars. A key point of difference is the notion that past events can be used to test and explain formalized models of individual behaviour. In *Governing the Commons*, Elinor Ostrom (1990) argues that individuals will negotiate new institutional arrangements when the (perceived) benefits of change exceed the costs of maintaining the status quo (see also Ostrom *et al.*, 1993; North 1990).[22] Central here is the notion that negotiating, monitoring, enforcing and re-negotiating rules of conduct entail particular costs, which individuals can ascertain in a rational manner. Institutional change is more likely, Ostrom (1990) asserts, when individuals share the understanding that failure to change will affect all individuals equally and negatively (also see White and Runge 1995). Unless the costs of inaction are readily apparent, however, individuals tend to discount their personal stake in the equation, making this type of interdependence highly elusive (Ostrom 1990; Pearce and Warford 1993).

Ostrom (1990) makes the case that rules are more likely to be followed and collective action achieved when individuals enjoy, and ideally institutionalize, a legacy of successful cooperation (cf. Putnam 1993). The logic here is twofold. First, it is argued, reciprocity creates an expectation (transmitted directly or through reputation) that individuals will be willing to cooperate again. Second, prior commitments provide a cumulative resource through which eligible 'participants' may forge more ambitious (and potentially beneficial) arrangements (cf. Putnam 1993; Ostrom *et al.* 1994a). As Ostrom argues,

> The investment in institutional change was not made in a single step. Rather, the process of institutional change in all basins involved many small steps that had low initial costs. Rarely was it necessary for participants to move simultaneously without knowing what others were doing. *Because the process was incremental and sequential and early successes were achieved, intermediate benefits from the initial investments were realized before anyone needed to make larger investments.* Each institutional change transformed the structure of incentives within which future strategic decisions would be made.
>
> (Ostrom 1990: 137; emphasis mine)

Framed in this way, past decisions, cooperation and institutional arrangements constitute forms of 'social capital' through which members can generate and obtain goods that would be difficult or impossible to obtain in isolation, or without this particular legacy. Similar historiographical approaches can be found in the behavioural studies of collective action in simulated commons dilemmas (e.g. Ostrom 1998; Ostrom *et al.* 1994a; Blomquist *et al.* 1994).

Although rooted in models of individual decision making and rational choice, Wade's treatment (1988) is more generally dependent on systematic variations in ecological variability and risk. At the heart of his thesis is the notion that rules emerge or evolve when the existing institutional arrangement is unable to cope with the relative weight of ecology, population, preferences and technology (or the lack thereof) (cf. Baland and Platteau 1996: Chapters 11–13). Drawing upon cross-sectional data collected from village research in South India, Wade (1988) makes the case that rules regulating grazing and common irrigation were more likely when population and other ecological pressures created a situation in which the probability of losses, resulting from malfeasance or free riding, was unacceptably high. Thus, soils with relatively high levels of water retention create good conditions for grazing which, in turn, attract a (hypothetically) unsustainable number of herders, thereby necessitating the creation of rules regulating access (Wade 1988). Once again, although Wade (1988) certainly draws upon archival material, and gives a picture of historical change in the villages, his assumptions, theory and methodology are primarily dependent upon a historiography, which discounts the importance of historical narrative (cf.

Mosse 1997) and uses past events to infer general and predictive theories about the conditions under which collective action will occur. This predilection stems in large part from the ahistorical way in which he frames his interest in the conditions under which:

> ... some peasant villagers in one part of India act collectively to provide goods and services which they all need and cannot provide for themselves individually,
>
> (Wade 1988: 1)

To summarize thus far, the works described here as neo-classical interpretations of collective action share an approach to a 'commons dilemma,' which stems from imperfect information and the inefficient distribution of cost, and solutions to this problem (rules that would govern access and encourage collective action to conserve). Moreover, their methodology and epistemology tend to be informed by a historiography that uses the past to infer general and predictive propositions about social behaviour.

Inequality, efficiency and the commons

A second point of commonality among the new institutionalists is the normative position they take on questions relating to inequality, poverty and the commons. In contrast to the entitlement school, which is centrally concerned with more 'traditional' questions of inequality, poverty and the commons, collective action scholars emphasize institutions and property rights in terms of their ability to achieve or explain the efficiency and stability of common property regimes. The prioritization of efficiency and equilibrium can be traced to the influence of neo-classical economics (see Chapter One), and to a fairly widely held assumption that improvements in common property management will produce social improvements for those whose membership confers rights of access and utilization in local CPRs (an assumption we explore critically below).

In *Governing the Commons*, Ostrom makes the case that Leviathan and privatization may not be the most efficient means of 'governing and managing diverse CPRs for which at least some potential beneficiaries cannot be excluded' (Ostrom 1990: 22). However, at no point in her analysis does she consider the kinds of burdens rules of common property – in particular rules designed to discourage free riding – would impose on those most adversely affected by the institutional arrangement. Similarly, Baland and Platteau (1996: Chapter 10) assess the management practices of 'pre-capitalist' societies by establishing whether such communities were able to control the size of their (human and non-human) populations, and whether they did so 'through the operation of ecologically oriented motives' (1996: 187). However, at no point in their analysis do they consider the moral arguments that have been levelled against the case for population control (Ross 1998); nor do they entertain the possibility that high fertility rates are not merely the result of individual decisions but also a response to the risks and incentive structures created by allegedly 'exogenous factors' such as industrialization and market capitalism (Ross 1998; Goldman 1998). [23]

If issues of inequality are incorporated at all, they are generally used to understand the ways in which social structure affects participation and collective action. Baland and Platteau (1996), for instance, consider the ways in which 'dominant actors' affect collective action to conserve natural resources. An important proposition they make is that effective collective action 'requires' an optimal degree of social inequality.[24] This can entail both the means by which individuals participate in economic affairs (what they call 'endowments') and the interests they take in this particular activity (Baland and Platteau 1996). Providing that dominant actors require the cooperation of a non-elite to achieve a collective good that is in their interest (e.g. efficient irrigation), inequality can provide the incentives, leadership and authority structures which are necessary for collective action:

> ... the elite have a disproportionately great interest in the effective regulation of water resources and this helps ensure that the required organization is started and effectively run.
>
> (Baland and Platteau 1996: 307)

Such assertions are principally based on the findings of Wade's (1988) study of irrigation management in South India, in which a strong association is made between *certain forms* of inequality and the collective provision of public goods. In *Village Republics*, Wade (1988) makes the case that villagers in Kottapalle were more likely to organize institutions governing irrigation and rights of grazing when the benefits of action outweighed the costs, and where the risks (of inaction) were high (cf. Ostrom 1990). Critically, calculations of this kind were more likely to be acted upon when sub-groups within a community were sufficiently small (cf. Olson 1965), interested and powerful enough to make a difference. As Wade states,

> Since benefits of corporate organization are positively related to land area, the claims that these households can make are sufficiently large for some of them to be motivated to pay a major share of the organizational costs. Debate and compromise are easier in a sub-group of this kind than in a larger and more heterogeneous group with more diverse preferences (cf. Ostrom 1990); so consensus about levels of provision is more readily reached.
>
> (Wade 1988: 190; parentheses added)

For Wade (1988), the key variables that explain the collective provision of common property arrangements are cost/benefit ratios and risk:

> Where the material benefit/cost ratio of field guards and common irrigators is high, it is likely that these services will be provided even in villages marked by a high degree of elite inequality ... On the other hand, where the benefit/cost ratio of field guards and common irrigators is low ... then whether the structure

of power and wealth is relatively equal or unequal makes little difference: public goods are unlikely to be provided.

(Wade 1988: 156–7)

A key point here is that social structure (and the relative distribution of costs and benefits within this structure) will have a strong bearing on whether collective action will be pursued and achieved. In *Governing the Commons*, Ostrom (1990) argues that individuals will change the rules of the game, providing they understand the costs and benefits of these decisions and (crucially) they have the opportunity to decide. Such opportunities are facilitated greatly by the existence of eight design principles: (1) clear resource boundaries; (2) clear rules of membership; (3) congruence between rules of provision/appropriation and local conditions; (4) arenas for 'collective choice'; (5) mutual monitoring; (6) 'graduated' sanctions; (7) mechanisms for conflict resolution; and (8) a state that is willing to recognize (or at least not challenge) local rights of organization (Ostrom 1990: 90).

Poverty, inequality and the commons: 'entitlement approaches'

Entitlement approaches differ from collective action approaches in three important ways. First, socio-economic equality and poverty reduction, as opposed to efficiency and the health of the commons, constitute major normative concerns. Second, rules are important in so far as they enhance, not restrict, access to the commons. Finally, the entitlement literature tends to favour a structural-historical approach, in which property rights and relations are contingent upon historically and contextually specific forces of social change (Table 2.1).

In contrast to the collective action school, the entitlement literature is *centrally concerned* with the problem of inequality, and with the ways in which formal and informal rules create and reinforce unequal access to common pool resources.

Table 2.1 Collective action and entitlement scholarship on the commons[25]

Body of scholarship	Collective action	Entitlement
Problematic	Conservation and management of common pool resources	Inequality and access in common property regimes
Conditions of change	Shifts in costs and benefits (Ostrom 1990), and in relative distribution of risk (Wade 1998)	Exogenous changes in population, markets and technology; structural change; state-enclosure
Mechanisms of change	Interest-based struggle, conflict, bargaining, negotiation among groups and individuals	Interest-based struggle, conflict and privatization of commons
Epistemology and methodology	Positive deductive search for general theory; methodological individualism	Inductive and deductive search for historically specific explanation or theory; structural historical

Implicit (and often explicit) in the entitlement literature is the normative assertion that socio-economic equality or, at least, a reduction in poverty, is desirable. Characterized in this way, common pool resources with 'unstable and low production levels' (Scoones 1996: 3) tend to be extremely important to marginal groups, such as women, ethnic minorities, the landless and the poor.[26] Seminal contributions in this literature include the work of N. S. Jodha (summarized in Jodha 2001), Blaikie and Brookfield (1987); Blaikie (1989), Mosse (1997), Ribot (1998), Beck (1994), Blair (1996), Leach *et al.* (1999), Goldman (1998), Prakash (1998), Cleaver (2000), Swift (1994), Scoones (1996) and Mearns (1996).

Although the benefits of risk minimization and the provision of public goods (such as common irrigation, etc.) are certainly valued within this more critical literature, the assumption that these benefits would necessarily 'trickle down' to the rural poor is a matter of some debate.[27] Taking the case of India (a country with large numbers of common pool resources, poor people and, unsurprisingly, studies of common pool resources and poverty), field studies have shown that common pool resources play an important role in sustaining the livelihoods of the rural poor. Beck and Nesmith, for instance, estimate that,

> (CPRs) currently contribute some US $5 billion a year to the incomes of poor rural households in India, or about 12% to household income of poor rural households.
>
> (Beck and Nesmith 2000 119)

These calculations are extrapolated from Jodha's 4-year study of poverty and common property in dryland India (summarized in Jodha 2001). Conducted between 1982 and 1986, Jodha's field studies cover 82 villages 'in seven major states in the dry tropical zone of India,' (Jodha 2001: 121). By measuring the resources and incomes provided by various types of commons (including community forests, irrigation reservoirs and pasture 'wasteland'), Jodha (2001) makes a convincing case that the rural poor are disproportionately dependent on the 'low pay-off options' offered by common pool resources. This, he argues, is due to low levels of competition with more affluent households, whose livelihoods are not dependent on local common pool resources. Such findings are consistent with other studies of common pool resources, livelihoods and poverty in India (e.g. Beck 1994) and in other parts of the developing world, which have demonstrated a vital link between local common pool resources and the livelihoods of the rural poor (see especially Beck and Nesmith's comprehensive review of common pool resources and rural poverty in India and West Africa, 2000).

Ethnographic studies of local commons and rural poverty have shown that although common property regimes may help to reduce the risk of resource depletion, assumptions that local institutions will be rooted in moral economies based on equity, welfare and social security are by no means guaranteed (Li 1996; Mosse 1997; Wade 1988). In his study of tank irrigation in Southern India, for instance, Mosse (1997) illustrates the ways in which common property institutions were manipulated to serve primarily the interests of male, high-caste members:

> … while the social control over access to tank water does indeed serve a collective need to manage risk and protect subsistence livelihoods, this does not mean that these systems are sustained by a community moral ethic. Social controls on the water commons may just as easily serve to establish relations of dominance and control …
>
> (Mosse 1997: 481)[28]

Reviewing a comprehensive amount of literature on poverty and the commons in India and West Africa, Beck and Nesmith (2000) argue that common pool resources provide substantial benefits to the rural poor, even when these systems are based on unequal social relations (cf. Baland and Platteau 1996). Moreover, like Jodha (2001), they observe that the benefits being provided by local commons in both regions have been undermined by the privatization of common property regimes, the commodification of resources provided by these systems and the exclusion of marginal groups, particularly women and children, whose labour is often highly dependent on the ability to access local resources (Beck and Nesmith 2000).

In my own research (Johnson 2000, 2001), which documents the establishment and enforcement of common property in Thailand's inshore fisheries, I found that households at the lowest end of the socio-economic spectrum experienced minimal improvement as a result of collective efforts to prevent trawlers and push nets from entering and depleting a local fishery. For those with poor and unproductive technologies, the returns that accrued from the resource (larger and more numerous species of fish) were tempered by their weak capacity and poor terms of trade. However, even for the relatively well-endowed owner-operators, the benefits that accrued from common property were barely keeping pace with the escalating costs of coastal fishing and variable returns. Indeed, the only households who appeared to be getting ahead in the industry were those with the capital, influence and contacts to command new or vital market niches, such as contract processing and direct marketing.

Such findings echo Ribot's conclusions (1998) about the relationship between market power and market access in Senegal's charcoal commodity chain:

> While local control may increase security, and perhaps the desire to maintain the resource base, it may not provide the economic means needed to do so …
>
> (Ribot 1998: 336)

Along similar lines, Leach, Mearns and Scoones (1999) use the example of *Marantaceae* plants (the leaves of which are used for wrapping food) in Southern Ghana to illustrate the ways in which the *type* of property regime can affect one's entitlement to a resource. If leaves are collected from forest reserve areas, women (who dominate the activity) must first obtain an official permit from the

Forest Department, which entitles them to collect the leaves. Off the reserve, collection rights can either depend upon the common property membership rules of the village (and its figures of authority) or individual arrangements with landholding families.

What this suggests is that membership in common property regimes – and the rights and duties these entail – are not sufficient determinants of livelihood outcomes. Equally important are the entitlements through which individuals obtain access to common pool resources and, crucially, the markets in which these commodities obtain value. Such insights are important, not least because they force an assessment of the historical factors, relations and processes that determine access to and control over local resources. Indeed, the entitlement literature is replete with cases of groups using the state and other forms of authority to recognise and enforce their claim over natural resources. For instance, Kurien (1992) describes the ways in which fishworker unions in Kerala used fasting and roadblocks to pressure the state to legislate against trawler companies. Similar descriptions are made about environmental struggles in Amazonia (Diegues 1998), forest battles in Cameroon (Nguiffo 1998), land rights in Namibia (Devereux 1996), peasant politics in Mexico (Fox 1996) and community rights in Central Sulawesi (Li 1996).

Entitlement scholars argue that it is against this backdrop of resource scarcity and competing forms of authority that collective action becomes important.[29] Framed in this way, the privatization of access rights is not only a consequence of individual choice or government policy on the commercialization of forest land (although the latter was, of course, relevant), but also a result of struggles to obtain and legitimate new forms of entitlement (cf. Johnson and Forsyth 2002).

Privatizing the commons: rights, incentives and rational choice

In the collective action literature, privatization of the commons is often presented as the result of misguided policy or rational choice, which individuals or 'decision makers' have chosen to implement. For instance, Ostrom (1990) and Baland and Platteau (1996) mobilize strong arguments against policies that would privatize the commons, or subject them to a system of over-centralized bureaucratic control. Such assertions are generally based on historical instances in which common property arrangements have been selected as the most viable means of resolving collective problems of access and restraint in common pool resources.[30] Grounded within a choice theoretic, the implication here is that different property rights regimes (e.g. common, private or a combination of the two) are the result of a selection process, in which resource users and resource managers impose institutions that best meet the needs of resource users and the resource base (Baland and Platteau 1996; Ostrom 1990; Wade 1988).[31]

Alternative interpretations understand the privatization of rights over natural resources as a historical process in which 'the state' is one of many interests whose power, autonomy and relationship with societal interests determine the allocation and functioning of property rights regimes (Li 1996; Mosse 1997;

Vandergeest and Peluso 1995). In contrast to the neo-classical school, which explains commons dilemmas (and their resolution) in terms of individually calculated responses to structural incentives (e.g. Wade 1988; Ostrom 1990), entitlement scholars understand and explain the degradation of natural resources in terms of a historical process grounded in the privatization and commercialization of local resource systems. Beck and Nesmith (2000), for instance, argue that the privatization and commercialization of local resources have eliminated poor people's access to the commons in rural India and West Africa, and at the same time undermined traditional indigenous practices of resource management. Along similar lines, Jodha (2001) argues that the failure to manage or rehabilitate common pool resources in rural areas of Rajasthan is due to a combination of government policy, which has encouraged privatization, indifferent leadership at the local level and low use-values, which reduces incentive to invest in their management.

Responding in part to the work of Jodha, Prakash (1998) argues that the 'increasing penetration of the modern market economy' has not only undermined local resource management structures but has also 'blurred and diffused' the relationships that underlie local institutions and traditions:

> The result is that the poor must largely fend for themselves within new social structures and private property regimes that, despite the occasional success of land reform policies, give them a marginal status ...
>
> (Prakash 1998: 175)

Challenging explanations that attribute excessive value to the calculations of individuals (e.g. Ostrom 1990) or to ecological structure (e.g. Wade 1988), David Mosse (1997) argues that the irrigation institutions he documented in Tamil Nadu reflected both the ecological characteristics and historical processes that gave rise to particular forms of institutional arrangement. The reason historical processes undermined water management institutions in the primarily clay soil lower catchment villages and not in the sandy soil upper catchment villages he studied was not (only) because life was inherently more risky in the latter (cf. Wade 1988), but because of specific historical processes in which widespread diversification out of irrigated rice cultivation and the in-migration of successful dryland cultivators had the (totally unintentional) effect of undermining a centuries-old system of irrigation management in the clay soil villages. The insight he draws from this analysis is that,

> ... the erosion of 'public' systems of resource use in black cotton soil villages, and their retention in sandy soil villages is the result of complex factors of settlement, caste power, property rights and economic opportunities, revenue systems and colonial markets. *This complexity has to challenge any oversimplified ecological deterministic reading of the pattern of collective action.*
>
> (Mosse 1997: 495; emphasis mine)

Along similar lines, Ian Scoones (1999) argues that institutional approaches to collective action and conservation of common pool resources (and here he includes the work of Ostrom 1990 and Bromley 1992) tend towards a 'balance of nature,' in which ecological and institutional processes are assumed to approach a state of equilibrium. Drawing upon theories of non-linearity, uncertainty and chaos, Scoones (1999) argues that 'new ecological' approaches have demonstrated the limitations of the equilibrium model, concluding that interdisciplinary approaches may help scholars to transcend the 'balance of nature view that has dominated both academic and policy discussions in the past,' (Scoones 1999: 496–97).

Such insights illustrate a number of important points about the tensions that have emerged in contemporary commons scholarship. First, there is clearly a tension between a scholarship grounded primarily in historical explanation of the kind that Mosse (1997) favours and one based on formal modelling of individual decision making and rational choice (cf. Goldman 1998). Although the lines are by no means neatly drawn, entitlement scholarship on the commons tends to favour a 'sociological-historical' method (Mosse 1997), which contrasts with the neo-classical assumptions that underlie the collective action school. Second, there is a *normative tension* within the literature, between a body of scholarship that is normatively pre-disposed towards the study of efficiency and conservation and one for whom inequality, poverty and exclusion constitute a central focus.

In summary, there is a strong body of criticism that challenges the collective action literature for its instrumental and historically de-contextualized understanding of common property relations. In critical response, proponents of this viewpoint have called for a deeper understanding of the ecological and socio-economic factors that affect the myriad relations on which property, common property and other forms of resource entitlement are based.

In the following section I argue that although demands of this kind are certainly justifiable, their appeals are unlikely to have a significant effect on the intellectual mainstream of commons scholarship. Specifically, I argue that the voice of this critical literature has been muted by two 'structural' factors: first, the entitlement and collective action scholarships are largely isolated from one another; second, the collective action school aims to contribute to an empirically grounded theory of social action, in which historical contingencies are merged with a scientific frame which is capable of testing falsifiable propositions about human behaviour. It is the challenges and implications of achieving this marriage to which we turn next.

Ships in the night: history and science in commons scholarship

In an important and provocative article, Arun Agrawal (2001) has made the case that commons scholarship has tended to be disproportionately focussed on case studies of small and local resource user groups, and that this preoccupation understates the important ways in which resource system attributes, user group attributes and the 'external social, physical and institutional environment affect institutional durability and long-term management at the local level' (Agrawal 2001: 1650–51). Methodologically, he argues that 'existing work has yet to develop fully a theory of what makes for sustainable common-pool resource management' (Agrawal 2001: 1651). Reviewing principally the work of Ostrom (1990), Wade (1988) and Baland and Platteau (1996), Agrawal suggests that:

> Systematic tests of the relative importance of factors important to sustainability, equity, or efficiency of commons are relatively uncommon … Also uncommon are studies that connect the different variables they identify in causal chains or propose plausible causal mechanisms. Problems of incomplete model specification and omitted variables in hypothesis testing are widespread in the literature on common property.
>
> (Agrawal 2001: 1651)[32]

Particularly undeveloped in the literature, Agrawal argues, are studies which isolate the causal importance of market forces, technology, population pressures and the state.[33] He makes a strong case that a more valid and predictive understanding of what makes for durable common property regimes will require a more systematic and scientific approach. This science will entail a 'careful attention to research design, index construction to reduce the number of variables in a given analysis, and a shift toward comparative rather than case study analysis' (Agrawal 2001: 1665). It will also require a larger number of cases and a systematic isolation of causal variables.

Agrawal's approach (2001) to common property research is firmly rooted in a positivism that seeks to construct general and predictive theories about the durability of common property regimes. This predilection is particularly apparent in his critique of the lack of research design in localized case study research:

> When a large number of causal variables potentially affect outcomes, the absence of careful research design that controls for factors that are not the subject of investigation makes it almost impossible to be sure that the observed differences in outcomes are indeed a result of hypothesized causes … If the researcher does not explicitly take into account the relevant variables that might affect success, then the number of selected cases must be (much) larger than the number of variables … *many of the existing works on the management of common-pool resources, especially those conducted as case studies or those that base their conclusions on a very small number of cases, suffer from the problem that they do not specify carefully or explicitly causal model they are testing.*
>
> (Agrawal 2001: 1661; parentheses in original, emphasis mine)

Therefore, underlying Agrawal's critique (2001) is an assumption that the goal of social science research on the commons is to produce general and predictive theories of human behaviour. However, at no point in the article does he defend the claim that social science work on the commons *should* aim to develop general and predictive theories. Going by the very vocal arguments that have been marshalled against the pursuit of de-contextualized theory we can infer that many interpretive commons scholars are not interested in developing general and predictive theories of human behaviour. Indeed, their ambitions are more modest and historically specific.

This divergence would not matter (i.e. there would not be a 'tension') were it not for the fact that both bodies of scholarship are centrally interested in low-income areas of the developing world. Reflecting on their findings from fieldwork in Zimbabwe, Campbell *et al.* (2001) highlight what they perceive to be a 'lack of congruence between the optimism of the (common property) literature and the realities on the ground' (2001: 595). This they attribute to one of two possibilities: (1) that the exceptionally inhospitable climate they find in Zimbabwe (the focus of their work) is substantially unique; or (2) that 'the optimism of the CPR literature (is) an artefact of particular ideologies or to the overstatement of the successes,' (Campbell *et al.* 2001: 595). The authors provide little detail as to what these 'particular ideologies' would be and how they would bias the work of the common property literature. However, they do highlight a concern that the common property literature's pre-occupation with formal modelling and theory building implies an attempt to engineer the institutions of societies in which most commons scholars are not members (cf. Goldman 1998). In critical response, they call for a 'more detailed ... (and) in-depth understanding of the processes involved in the evolution and dynamics of institutions' (Campbell *et al.* 2001: 596).

Such concerns are echoed by Prakash, who argues that the commons literature (and here he is referring to what I have called the 'collective action school'),

> ... has largely circumvented the implications of internal differentiation ... the plurality of beliefs, norms and interests ... the effects of complex variations in culture and society, as well as social, political and economic conflict relating to the commons.
>
> (Prakash 1998: 168)

The problem, Prakash argues, is that the policy analyst's 'abstraction from the complexities of field settings' may lead to 'a reification of concepts, models and strategies' (Prakash 1998: 168). Similar assertions are made by Mosse, who argues that common pool resource management 'cannot (as is so often the case) be isolated from context and viewed as a distinctive type of economic activity' (Mosse 1997: 486). Earlier in the article, and in more detail, he argues that,

> ... an institutional analysis of indigenous resources systems is unlikely to be useful unless it has first correctly characterized the social relations and categories of meaning and value in a particular resource system. In the first place this means resisting the tendency to impose a narrow definition of economic interest, utility and value ...
>
> (Mosse 1997: 472)

Perhaps most critical of the neo-classical literature is Michael Goldman, who identifies

> ... a fundamental tension between knowledge production and historical consciousness, a tension between casting a blind eye towards the destructive forces of capitalist expansion onto the commons and a broad smile that beams at the 'underskilled' local commoner who defies all odds by protecting the commons.
>
> (Goldman 1998: 21)

Coming back to the tensions that underlie contemporary commons scholarship, it is not at all clear how differences about foundational normative principles or problems (such as the existence of inequality) may be overcome. Central to the tensions that exist in the commons literature is a divergence between a social science, which seeks to build theory on the basis of scientific empiricism, and an ethnography, which rejects the universalism that underlies the scientific approach. The problem this creates for the study of social phenomena (such as common property relations) is that proponents of context and ethnographic method and those favouring a scientific frame are left with little common ground on which to stand. This is mostly due to the very large gap that exists between the historicist rejection of universal theorizing, which appears fundamental to the critique of critical commons scholars, and the central assumptions of positivist social science.

In response to all of this one may argue that normative and methodological differences are perhaps of less importance than the ways in which disciplines and disciplinary practices accommodate different meta-theoretical aims and values. Although the journals (e.g. *World Development, Development and Change, Politics and Society*) and scholarly institutions (e.g. the International Association for the Study of Common Property [IASCP]) in which research and debate about the commons tend to appear most regularly are certainly not stifling debate about these issues – on the contrary, it is in these forums that we find the tensions thus described – one cannot ignore the fact that collective action and entitlement scholars are working at (normative and methodological) cross-purposes, and that they tend to

Table 2.2 Conservation and entitlement in commons scholarship[34]

Combinations and number of citations	
'Tragedy' and the commons	468
'Conservation' and the commons	355
'Efficiency' and the commons	111
'Poverty' and the commons	105
'Exclusion' and the commons	34
'Moral economy' and the commons	9 (three by the same author)
'Entitlement' and the commons	6
'Sen' and the commons	9 (one authored by Sen)
'Ostrom' and the commons	270
'Sen' and entitlement	4

be working separately. A search on the IASCP's comprehensive search engine, for instance, produced a large number of commons studies whose key words included 'tragedy,' 'conservation,' 'efficiency' and 'Ostrom,' but relatively few describing 'exclusion,' 'entitlement' or 'Sen' (see Table 2.2).

Granted, an exercise of this kind reveals little about the explicit content of the papers, articles and books contained in the library. Moreover, the data set is limited to one search engine! That being said, the search engine is managed by an organization that is arguably at the forefront of commons scholarship and therefore represents an important articulation of where the science is and where it is going. Moreover, one could make the case that the selection of keywords and titles reveals a great deal about the discourses and debates in which scholars wish to be engaged. Prevarications in hand, I would venture to conclude that the very low number of articles dealing with Sen, entitlement and exclusion is at least consistent with my earlier assertion that those aiming to understand the efficiency and conservation of CPR institutions (apparently a central focus of the IASCP) are generally not the same as those who are interested in issues of entitlement, inequality and exclusion.

Concluding remarks

This chapter has explored the meta-theoretical differences and debates between new institutional and historical approaches to the study of poverty, livelihoods and the commons. A central claim was that the meta-theoretical differences that divide historical and positivist approaches to the study of poverty, livelihoods and the commons are essentially differences between a social science, which seeks to build theory on the basis of scientific empiricism, and an ethnography, which rejects the universalism that underlies the scientific approach. For Agrawal (2001), 'context' is a variable that has bearing on explanation, but one that sits outside of the cultural processes and practices that inform the construction of research questions, hypotheses and methods. From an interpretive perspective, the kinds of positivism being advanced by Agrawal and Ostrom simplify or ignore entirely historical details whose appreciation is essential for understanding social phenomena, such as the creation and privatization of property rights and relations. In so doing, they raise a number of fundamental questions about the nature and study of social phenomena.

What makes neo-classical theorizing about collective action and the commons so striking is the relative lack of normative theorizing about the philosophical and material implications of constructing a theory based on principles of efficiency and exclusion. Although there is certainly no lack of empirical work on the ways in which people value and understand the commons, and how this has changed over time, a priori justifications for the establishment and maintenance of common property arrangements are almost exclusively based on the neo-classical assumption that common property arrangements can yield outcomes that are at least as efficient or more efficient than ones based on private property or centralized bureaucratic control. Largely unquestioned in this literature are the core values on which different systems of property – and our understanding of these systems – should be based.[35]

The arguments and evidence considered in this chapter also suggest a divergence between a positivist social science, which seeks to build theory on the basis of scientific principles, and an ethnography, which rejects the universalism that underlies the scientific frame. The postmodern rejection of all things universal, and the impasse this created for development theory and practice, was certainly part of this process (see Chapter Three). However, equally important was an evolution in American political science in which critical reflections about core values appear to have been subsumed by a desire to emulate the central aims and assumptions of natural science. Biases of this kind were particularly pronounced during the behavioural movement of the post-war period. Nevertheless, as I have argued in this chapter, one can make the case that the retreat from historicism is equally (and in some ways more) apparent in the new institutionalism, a part of which is collective action scholarship on the commons.

3   Exporting the model

Marxism, postmodernism and development

The philosophers have only interpreted the world, in various ways; the point, however is to change it.

Karl Marx (*Theses on Feuerbach*)

The idea of development provided a way of narrating world history, but not necessarily a rationale for acting upon it.

Frederick Cooper and Randall Packard (1997: 7)

Classifications are the cornerstone in any theorizing.

Angkie Hoogvelt (2001: 217)

Introduction

The previous chapter explored the challenge of incorporating history into a neo-classical/positivist frame. This chapter now explores the dominant efforts within Marxism to develop a unified and predictive theory of capitalist development and social change. Its principal focus is on three central problems in Marxist social theory. One concerns the *classification* of economic and social relations. A second concerns the *interpretation* of historically contingent processes and events. A third concerns the methodological terms on which the people standing behind the structures of political economy may be *represented* in Marxist social theory.

In what follows I shall argue that Marxism has been deeply divided about the ways in which and the extent to which culturally contingent processes and institutions (e.g. dowry, religion, ethnicity, the caste system) may be theorized and explained in relation to an idealized understanding of capitalist transition. To make my case, I distinguish between two major traditions within Marxist social theory. The first, influenced by the goals and assumptions of nineteenth century social science, aims to identify the underlying trends and forces that explain the development of capitalism (and with it the contradictions that would lead to its ultimate demise). A second, influenced by a series of pivotal events that either contradicted or (at the very least) failed to conform to the theoretical assumptions advanced within Marxism, and began to embrace a more critical or 'hermeneutic' analysis of the ways in which local and locally contingent processes and events may affect class consciousness, class struggle and social change.[36]

The chapter proceeds with the following section, which first lays out the core ideas and assumptions that Marx used in an effort to understand and explain the development of capitalism in Western Europe. The next section then considers the 'dependency debate' concerning whether and to what extent European colonialism (and later the green revolution) had set in motion the kinds of investment and class differentiation that would approximate the capitalist transition. The third section explores the literature that emerged in the wake of the dependency debate, making the case that the retreat from ideology and 'grand' social theory in Marxism can be attributed to the postmodernism of Lyotard, Derrida, Foucault and 1968, but also to a much longer-standing struggle within Marxism to reconcile the lived experience of class conflict with an objective and universal theory of class struggle and social change. The concluding section reflects upon the wider implications on development theory and praxis.

Theorizing the transition: Marxism, dependency and (capitalist) development

As a student of economic history, Marx was centrally interested in the contradictions that underlie an economic system rooted historically in the separation of labour and capital. To explain this transition, Marx starts from the premise that all social relations – and therefore one's understanding of these relations – are rooted in the assumption that we sustain ourselves through nature and that we do so not in isolation, but with others. As Marx writes in *The German Ideology*,

> The first premise of all human history is, of course, the existence of all liv-
> ing human individuals. Thus the first fact to be established is the physical
> organization of these individuals and their consequent relation to the rest of
> nature.
>
> (*The German Ideology*, quoted in Freedman 1961: 3)

Framed in this way, the crucial mode (and the crucial transition) was the capitalist mode (and the transition to capitalism), in which 'the tie between producers and the means of production is severed for good' (Wolf 1997: 77).[37] The need to sell and the ability to purchase the power of labour therefore constitutes a crucial point of transformation between the 'capitalist mode' and any other mode that precedes it.

Like Hegel (see Box 3.1), Marx believed that history manifests itself in terms of identifiable epochs or stages, and that historical changes may occur as a result of 'dialectical' contradictions within society. Unlike Hegel, he felt that the ideational super-structure of society would change only as a result of political actions to resolve material relations and contradictions within society.[38] At the heart of this dialectic was a fundamental contradiction between the short-term needs of capital and the long-term wealth and productivity of capitalism.

Marx argued that capitalists were under constant pressure to increase surplus,

Although he shared with Marx the idea that social change would arise as a result of contradictions within society, Hegel's belief was that agency to change the structure of society was dependent upon the *zeitgeist* or moral and ideological spirit of the times. The implication was that any effort to change material relations within society (e.g. slavery or private property) would first require an ideational change to a different social order. For Hegel, material social relations were dependent on a 'totality' of ideas and values, which were historically specific to particular 'epochs' in time (Morrow and Brown 1994: 94).

Box 3.1 Theories of history: Hegel and Marx

either by extending the length of the working day or by intensifying through the use of technology (i.e. machinery) the value of labour. Machinery, Marx argued, was instrumental in the sense that it reduced the 'labour-time necessary for the labourer to produce his subsistence,' and it also enabled the use of 'those whose bodily development is incomplete,' especially women and children (*Capital, Volume I*, cited in Freedman 1961: 84–85). However, by 'throwing every member of (the) family onto the labour market,' machinery also depreciated the value of labour power and set in motion a series of forces which were, for Marx, the underlying contradictions of capitalism. Assuming that capitalism could only reproduce itself by extracting surplus value from labour, Marx theorized that the (constant) pressure to replace variable capital (labour) with constant capital (machinery) would eventually induce a falling rate of profit which in turn, would lead to economic crisis, social upheaval and revolutionary change. Capitalist accumulation therefore entailed a dialectic of conflict and contradiction, in which 'a fall in the rate of profit speeds the process of accumulation, and accumulation hastens the fall in the rate of profit,' (*Capital, Volume III*, cited in Freedman, 1961: 175).[39]

Exporting the model

For Marx, the social and economic changes that transpired between the sixteenth and nineteenth centuries in England created the necessary conditions under which the capitalist mode of production could arise. First, the English economy underwent a long series of innovations in agriculture and cottage industry, in which the traditional system of 'putting out' was gradually replaced by the mechanized and synchronized factory system of organized labour (Wolf 1982 [1997: Chapters 9 and 10]. Second, and part of this process, the enclosure of open and feudal land systems created a pool of labour whose only means of survival could be derived through the sale of wage labour. Third, the period witnessed a massive military-territorial expansion, which provided the English economy with new and unprecedented access to raw materials (e.g. gold, silver, spices, indigo, cotton) and eventually to overseas markets for English manufactures (Wolf 1982 [1997]).

However, discerning whether these properties were exhibiting themselves outside of the advanced industrialized economies of Western Europe and North America was a different matter, frustrated significantly by the fact that many potential entrants or participants in the urban and rural proletariat were engaged simultaneously in capitalist forms of wage labour and in feudal forms of peasant agriculture (Leys 1996; Roxborough 1979). Although Marx identified 'a number of different modes,' including 'an original, primitive, communitarian mode,' a 'slaveholding mode of classic European antiquity,' 'a Germanic mode, supposedly characteristic of the Germanic peoples in their early migrations,' 'a peasant mode,' 'a feudal mode' and 'an Asiatic mode,' (Wolf (1982) [1997]: 75), his writing did not theorize extensively about the conditions under which capitalism would arise outside of Western Europe.

Interpreting Marx's body of work, Lenin theorized that Russian agriculture would follow a particular yet historically determined *path* of agrarian transformation and class struggle (Harriss 1982; Lenin 1982; Bernstein 1994; Brass 1991; Byres 1983, 1991).[40] Central to this transformation was a process whereby capitalism emerges as the dominant mode of production in agriculture, creating an agrarian structure in which the peasantry is increasingly differentiated along the lines of an affluent class of a 'peasant bourgeoisie' and, at the other extreme, a stratum of poor peasants who are either absorbed into agricultural labouring or into the urban proletariat (Bernstein 1983; Byres 1983, 1991; Harriss 1982; Djurfeldt 1982; Lenin 1982). Within this wider spectrum of agrarian class relations, Lenin theorized that 'bondage, usury, labour-service' and other forms of money-lending would constrain the differentiation of agrarian classes and the development of capitalism (Lenin, cited in Harriss 1982). Framed in this way, the decline of a feudal agrarian order would entail the rise of the increasingly affluent capitalist farmer and a socially-progressive struggle against a feudal or semi-feudal landlord class (Brass 1991; Byres 1983, 1991).

Lenin therefore highlighted what was at the time (and still remains) a crucial means of organizing and extracting agricultural and rural labour. However, the notion that agrarian class relations would tend to polarize along the lines envisaged by Lenin was challenged (by Kautsky and Chayanov, among others) for suggesting that the development of capitalism would entail a class structure polarized along the lines of a land-owning bourgeoisie and a labouring proletariat. Kautsky, for instance, theorized that capitalist and smallholder 'peasant' agriculture could co-exist by binding peasant production to the power of the factory and of industrial production (Kautsky, cited in Harriss 1982). Chayanov took this even further by suggesting that the fate of small-holder agriculture varied not (only) with the spread of capitalism, but also with the cyclical and demographic variations of investment decisions made within the household (Chayanov cited in Djurfeldt 1982).[41]

Within development studies, the 'classic' debates between Lenin and his critics framed a number of long-standing debates about the ontological nature and empirical existence of capitalism outside of North America and Western Europe (cf. Roxborough 1979; Harriss 1982; 1994). Set in the context of the green revolution, many of these debates became deeply (and some would say irreversibly) divided about the extent to which the introduction of new agricultural technologies (i.e. high-yielding varieties of rice, wheat and maize, nitrogen-based fertilizers and pesticides) had set in motion the kinds of investment and class differentiation that would approximate the capitalist transition (Booth 1985; Harriss 1994). In so doing, they raised a number of questions about the ways in which different modes of production interact and 'articulate' with one another, and how this affects traditional agrarian institutions governing land, labour and capital. One important area of difference concerned the extent to which preferential access to land, credit and green revolution technologies would foster processes of commercialization and consolidation in agriculture that would create and sustain a large and dependent pool of agricultural wage labour. A second concerned the extent to which the processes being described in these instances approximated the 'capitalist' mode of production.

The dependency debate

Perhaps the most vitriolic of the debates that ensued within the broader Marxist paradigm was the so-called 'dependency debate,' a long series of essays that stemmed from the idea that capitalism in the former colonies was 'structurally constrained' by historical and unequal relations between the peripheral and primarily agricultural economies of Africa, Asia and Latin America and the advanced industrialized economies of North America and Western Europe.

The idea that Third World societies were historically dependent on the production processes and consumption preferences of the advanced industrialized economies reflected a number of trends that, by the 1960s, were becoming increasingly difficult to reconcile with Marxist and liberal theories of modernization and development. One was a recognition (observed coincidentally by Raul Prebisch and Hans Singer) that the performance of Third World economies was 'structurally constrained' by their dependence on raw material and agricultural exports and by the fact that demand for industrial goods is more elastic (i.e. it tends to be more responsive to changes in income) than is demand for primary goods (Peet and Hartwick 1999). The result was a downward trend in the price of commodities and in the terms of trade of primarily raw material exporting economies.

A second and related factor was a largely theoretical debate about whether capitalism could take form in societies affected and rendered dependent by European colonialism. Attempting to develop a predictive theory about the roots and nature of Third World dependency, Andre Gunder Frank (1969) argued that colonialism had altered permanently the internal class structures of Third World societies, rendering them necessarily dependent upon the incomes, markets and production processes of the core (whose definition was also a subject of major debate). Through unequal land ownership, debt relations and bonded labour, Frank (1969) argued, the peasantry was locked in a state that was neither capitalist nor feudal, providing labour in exchange for wages that perpetually failed to meet the needs of subsistence (forcing labour to 'reproduce' itself only through other forms of usury and servitude). Through foreign ownership and investment, multinational capital was able to extract and transfer the surplus of agricultural and commodity production from the periphery to the core. 'Underdevelopment,' in Frank's words, was not due to the survival of

> ... archaic institutions and the existence of capital shortage in regions that have remained isolated from the stream of world history. On the contrary, underdevelopment was and still is generated by the very same historical process which also generated economic development: the development of capitalism itself.
>
> (Frank 1969 [2000: 163])

Methodologically, Frank's interpretation (1969) was almost entirely dependent on trends and data he had extrapolated from the Latin American experience. Drawing upon case study material from Chile and Brazil, he suggested that de-industrialization and dependency were highest in the regions and periods during which ties with the core were strongest. Conversely, 'the satellites experience their greatest economic development and especially their most classically capitalist industrial development if and when their ties to their metropolis are weakest,' (Frank 1969 [2000: 164]). To develop a more general theory of dependency, he advanced a systematic program of research that would in theory isolate the historical instances during which ties between the core and periphery were at their weakest. Focussing on the world economic depressions of the late nineteenth and early twentieth centuries, he suggested that social scientists could observe empirically the economic implications of periods (and policies) in which peripheral economies were more or less connected to the core.

In retrospect, Frank's assertion that the class structures of Third World societies were locked in a state of semi-feudal/semi-capitalist dependency was in a number of ways deeply problematic. For one, it appeared to lack a theory of change (Peet and Hartwick 1999). Although it provided sound descriptions of the ways in which rural class structures extracted labour and rent from the rural periphery (especially in Latin America), Frank's emphasis on landed inequality appeared to underplay the possibility that these processes may also lead to accumulation and differentiation (Peet and Hartwick 1999). Second, the notion that foreign capital was inherently parasitic underplayed the possibility that closer ties with multinational capital may under certain circumstances lead to processes of technological innovation and industrial change (Cardoso 1979).

Perhaps the strongest (empirical) argument in favour of this position came from Bill Warren, whose posthumous study of capitalism and dependency (1980) suggested that European imperialism did not prevent or distort capitalism in the periphery but had, in fact, laid the foundations for capitalist development. Drawing largely upon macro-economic data, Warren (1980) argued that indicators of economic and social development (i.e. GNP per capita, income inequality, caloric intake, infant mortality, life expectancy and levels of industrialization) had all improved since the end of the Second World War. Moreover, he argued, these improvements had taken place in countries whose economies were closely tied to the core of European colonialism. Although he did not live to see the fate that would follow the international debt crisis and the structural adjustment experiments of the 1970s and 1980s (see below), Warren helped to articulate for many the difficulty neo-Marxist theories of dependency and underdevelopment had in explaining cases in which close relations with the core had led to substantive improvements in economic productivity and well-being.

Building upon Frank (1969), writers like Arghiri Emmanuel (1972) and Samir Amin (1974) thus began to theorize in much greater detail the historical conditions under which economic surplus was transferred from the periphery to the core. Emmanuel (1972), for instance, suggested that capital mobility, labour immobility and wage differentials (between the periphery and the core) fostered a process of unequal exchange whereby low-income countries became perpetually dependent on agricultural exports, whose labour content was high, and expensive capital goods imports, thereby depleting domestic savings and foreign reserves (Peet and Hartwick 1999). Drawing upon the Brazilian experience, Fernando Henrique Cardoso (1979) observed that economic dependency in Brazil was not a function (simply) of colonialism but, rather, of a 'triple alliance' among domestic capital, foreign capital and the nation state, an idea that would help to explain processes of industrialization and capitalist transition in Northeast Asia.

World systems theory

Perhaps the most ambitious of those working in the dependency paradigm was Immanuel Wallerstein (1974, 1979), whose 'world systems theory' suggested that the relative power of core, peripheral and 'semiperipheral' economies was contingent upon the ability of nation states to project and protect national economic interests through the use of military-industrial power. Controversially, Wallerstein argued that the entire world (including the developing world and the Soviet bloc) had by the end of the seventeenth century become part of a capitalist world system, suggesting that capitalism did not necessarily require the violent separation of labour and capital that Marx had theorized but, rather, simply 'production for sale in a market in which the object is to realize the maximum profit' (Wallerstein 1979 [2000: 197]). In greater detail, he explains the emergence of capitalism in Northwest Europe:

> By a series of accidents – historical, ecological, geographic – northwest Europe was better situated in the sixteenth century to diversify its agricultural specialization and add to it certain industries (such as textiles, shipbuilding, and metal wares) than were other parts of Europe. Northwest Europe emerged as the core area of this world-economy, specializing in agricultural production of higher skill levels, which favored (again for reasons too complex to develop) tenancy and wage labor as the modes of labor control. Eastern Europe and the Western Hemisphere became peripheral areas specializing in export of grains, bullion, wood, cotton, sugar – all of which favored the use of slavery and

coerced cash-crop labor as the modes of labor control. Mediterranean Europe emerged as the semiperipheral area of this world-economy specializing in high-cost industrial products …

(Wallerstein 1979 [2000: 199])

Wallerstein (1974, 1979) argued that capitalism began to 'stabilize' around 1640, displacing in the process other 'mini-systems' of 'simple agricultural or hunting and gathering societies' (Wallerstein 1979 [2000: 192]). 'Leaving aside the now defunct minisystems,' he argued, 'the only kind of social system is a world system, which we define quite simply as a unit with a single division of labor and multiple cultural systems' (Wallerstein 1979 [2000: 192]). Whether and how these different 'cultural systems' engaged or disengaged with the world system was contingent upon the ability of ethnically differentiated groups (what Wallerstein called 'ethno-nations') to achieve, through trade, trade protection, competition and war, an advantageous position within the world system:

> The functioning then of a capitalist world-economy requires that groups pursue their economic interests within a single world market while seeking to distort this market for their benefit by organizing to exert influence on states, some of which are far more powerful than others, but none of which controls the world market in its entirety.

(Wallerstein 1979 [2000: 202])

Therefore, Wallerstein (1974, 1979) advanced a theory which suggested that peripheral and semiperipheral economies could change their place in the world system through business cycles (arising primarily as a result of wage differentials and transnational movements of capital), war-making and strategic efforts to disengage from the capitalist world economy. Framed in this way, the Soviet Union and the People's Republic of China were both 'part of' the world capitalist system. However, the crucial point is that they had (for a time) successfully distorted the market as a means of consolidating their power in relation to the core economies of North America and Western Europe.

Although much can be said about Wallerstein's account of the historical conditions under which capitalism began to emerge in north-western Europe, our main concern at this point is with the way in which he classifies capitalism, and how this differs from classical Marxist theory. First, Wallerstein's assertion that capitalism began to stabilize 'around 1640' long pre-dates the period (of the eighteenth and nineteenth centuries) during which most Marxists would locate the emergence of an identifiable working class. Second, Wallerstein's suggestion (1979 [2000]) that capitalism implies 'production for sale in a market' appeared to underemphasize the crucial separation of labour and capital. For instance, Robert Brenner (1977) argued that Wallerstein (1974) and Frank (1969) advanced what amounted to a 'neo-Smithian' emphasis on trade and commerce, which in his view obscured the vital importance of class relations and production in the development of capitalism. For Brenner (1977), the relations of exchange that defined and determined relations between core, peripheral and semiperipheral regions and economies were of less importance than the specific historical factors that determined the structure of property rights in land, the (historical) creation of 'free' labour and (more generally) the transformation of productive forces from a feudal social order to a primarily capitalist system of farming and industry. Framed in this way, wage differentials, rates of investment, transfers of wealth, unequal exchange, etc., were all the living embodiment of historical processes (e.g. the English enclosures, peasant reforms in France, the lack thereof in Poland) that led to specific class formations, whose nature and productivity influenced 'the rise' and the nature of trade and investment.

Finally, Wallerstein's assertion that countries, regions and 'ethno-nations' may be understood primarily in terms of the 'function' they provide to the capitalist world system reduced, to apparent insignificance, the particularities and nuance of human agency, history and experience (Wolf 1997, and below). Take, for instance, his treatment of the 'semiperiphery:'

> … one might make a good case that the world-economy as an economy would function every bit as well without a semiperiphery. But it would be far less *politically* stable, for it would mean a polarized world-system. The existence of the third category means precisely that the upper stratum is not faced with the unified opposition of all the others because the middle stratum is both exploited and exploiter.
>
> (Wallerstein 1979 [2000: 201])

Reflecting upon this tendency to address questions of history and identity in relation to the imputed needs and functions of the capitalist world system, Eric Wolf argued that dependency theorists had become pre-occupied with efforts:

> … to understand how the core subjugated the periphery, and not to study the reactions of the micro-populations habitually investigated by anthropologists. Their choice of focus thus leads them to omit consideration of the range and variety of such populations, of their modes of existence before European expansion and the advent of capitalism, and of the manner in which these modes were penetrated, subordinated, destroyed, or absorbed, first by the growing market and subsequently by industrial capitalism.
>
> (Wolf 1982 [1997: 23])

Debates about colonialism, capitalism and dependency therefore revealed fundamental differences about what capitalism was (as an ontological category) and what (in theory) it could be. For dependency theorists, colonialism had altered permanently or necessarily (through plantations, slavery and other colonial modes of production) the internal class structures of Third World societies, thereby preventing the formation of a separate capitalist and working class. For traditional Marxists, capitalism was both possible and discernable in 'the periphery,' manifesting itself most significantly in the historical transformations (e.g. in Western Europe or in Northeast Asia) that separated either substantially or permanently the link between 'free' labour and the means of production (Roxborough 1979).

After dependency

By the end of the 1970s, the notion that Third World economies were perpetually locked in a trap of low-income dependency had become increasingly difficult to square with the empirical fact that the ex-colonial economies of South Korea and Taiwan had quite clearly experienced sustained processes of industrialization, capital accumulation, and socio-economic change (Harris 1987; Evans 1995; Kohli 2004). Whether these experiences approximated the factors highlighted within Marxism, the existence of industrialization in Northeast Asia inspired a new body of scholarship that began to theorize the ways in which the state could actively engineer the conditions for capitalist development (e.g. Johnson 1982; Amsden 1989; Wade 1990; cf. Rapley 2002). In so doing, it complemented the work of other 'late dependency' theorists (e.g. Cardoso 1979), whose analysis suggested that the existence of dependency or autonomy was contingent upon historically specific inter-relations among foreign capital, domestic capital and the nation state (Peet and Hartwick 1999).

In theory, the rise of the Asian newly industrializing countries (NICs) – and with it the apparent collapse of the dependency paradigm – should have reinvigorated the idea of explaining capitalism through the 'traditional' Marxist lens. However, the reverse, in fact, became true. Within political science and the international political economy (IPE), comparative perspectives on globalization and development began to highlight the varied ability of the Northeast Asian states to manage the entry of their economies into the world economy (e.g. Evans 1995; Weiss 1998; Kohli 2004). Particularly formative in this regard was the 'developmental state' literature (e.g. Johnson 1982; Amsden 1989; Wade 1990), which highlighted the ability of governments in South Korea, Taiwan and Japan to protect infant industries through the use and gradual removal of tariffs, directed credit and other economic subsidies.

Although they emphasized questions of power and political economy, the aims and assumptions of the developmental state literature were now a far cry from the kinds of politics being advanced by 'traditional' Marxism (Fine 2001; Hoogvelt 2001). First, the notion that economic development could be explained primarily in terms of class conflict and class relations was no longer central. Although the rise of the Asian NICs inspired newfound interest in the possibilities of 'late industrialization' (Amsden 1989), the central focus was primarily on state capacity and industrial policy, not on the organization (and suppression) of industrial labour (Deyo 1987; Fine 2001). Second, the intellectual and ideological terms of discourse had quite clearly shifted from one that emphasized the dialectical nature of class struggle to one that defined itself primarily in terms of 'market versus state' (Fine 2001: Chapter 8). Indeed, one could argue that the 'developmental state theory' of Johnson (1983), Amsden (1989) and Wade (1990) had more in common with the German theorist Friedrich List than it did with Engels or Marx (cf. Fine 2001; Chang 2002).[42]

A second body of scholarship was the 'regulation school,' an area of neo-Marxist scholarship that aimed to understand the social and historical conditions under which capital adapts to periodic crises and contradictions inherent to the capitalist system (Hoogvelt 2001). Central to the work of Michel Aglietta, Alain Lipietz and Charles Boyer (all reviewed in Hoogvelt 2001: 115–17) was the idea that capitalism operates and adapts in relation to a 'regime of accumulation,' which constitutes a 'relatively stable and reproducible relationship between production and consumption defined at the level of the international economy as a whole,' (Hoogvelt 2001: 116). One important area of research concerned the perceived shift from 'Fordist' to post-Fordist regimes, in which stable, site-specific modes of production and consumption (upheld by Keynesian policies of full employment and public investment) had been replaced by mobile, temporary arrangements in which outsourcing, just-in-time practices and increasingly flexible labour arrangements facilitated a new international division of labour within the world economy (Harvey 1990; Peet and Hartwick 1999; Hoogvelt 2001).

Whether these processes would lead to industrialization and capital accumulation, the regulation school generated considerable discussion and debate about the nature and extent of real capital formation in the developing world, fostering new theoretical insights about the social organization of markets, commodity chains and capitalism writ large (Hoogvelt 2001; McMichael 2004). However, in contrast to the 'traditional' focus of classical Marxism on revolutionary action and class struggle, the regulation school appeared far more inclined to document and describe the conditions under which capitalism could reproduce its existence than to search for instances in which capital accumulation and class differentiation may lead to revolutionary change. On the 'regulation school,' for instance, Angkie Hoogvelt suggests:

> There is no conception of social progress; no eschatological belief in the forward march of history; no political commitment to surrender the freedom of the intellect to a course that history has charted.
>
> (Hoogvelt 2001: 115)

Rather, she argues,

> There are no certain outcomes predetermined by inherent tendencies. What the new mode looks like is entirely contingent, both historically and nationally. It depends on the outcomes of specific, local, social and political struggles, strategies and compromises, and the pre-existing local institutional context ... Hence its research agenda has been focused on precise, detailed and empirical analyses of the content and the actual contingent movement of capital, which is so diverse in its manifestations that it leaves considerable scope for historical and national variation.
>
> (Hoogvelt 2001: 116)

In short, the 'Marxism' that emerged in the aftermath of the dependency debate was in many ways a shadow of its former self. To understand the retreat from grand social theory, the following section explores two bodies of criticism that began to question the idea that history could be used to understand and advance a universal and predictive model of social change. One criticism concerns the interpretation of historically contingent processes and events while the other concerns the representation of individual worldviews and experiences in Marxist social theory.

'Hermeneutic Marxism': problems of agency, identity and 'alienation'

By the end of the nineteenth century, the notion that Marxism could be used to understand the diversity and complexity of human experience had become the object of considerable critical scrutiny. One early and important figure in this regard was Ludwig Wittgenstein (1889–1951), whose 'linguistic philosophy' suggested that the desire of Marxism and especially positivism to develop a pure and formal representation of reality understated the ontological challenge of connecting objective or universal words and concepts with inter-subjective meanings and experiences (Morrow and Brown 1994). Another important figure was John Austin (1911–1960), who argued that the ability of language to confer meaning was dependent upon the social and cultural context in which different 'speech acts' about the nature of reality and of discourse were made (Morrow and Brown 1994). Ian Shapiro captures the essence of this important idea:

> Social reality is linguistic reality on this view. When human beings do things like create obligations or social contracts they do this through language, not by some other means that is then described by language. Understanding reality means understanding the linguistic processes that give rise to it.
>
> (Shapiro 2005: 5)

Doubts about the scientific claims and assumptions of positivism and Marxism also led to more critical forms of inquiry about the nature of human consciousness and social action. Edmund Husserl (1859–1938), for instance, rejected the idea that positivism could represent subjective realities in objective terms, and sought instead to uncover the hidden meaning that individuals associate with social practices (Morrow and Brown 1994). Husserl's work became known as *phenomenology*, which emphasized 'the idea that human subjects are formed by the historical cultural practices in which they develop' (Dreyfus and Rabinow 1983: xxi). A related field was *hermeneutics*, pioneered by Husserl's student Martin Heidegger (1889–1976), which suggests that subjective realities may be captured by uncovering the 'lived experience' and everyday practices of social life.

Early forms of linguistic philosophy, phenomenology and hermeneutics therefore challenged Marxism, positivism and other forms of nineteenth century social science for the largely uncritical ways in which they separated facts and the collection of facts from their social and cultural context. Where phenomenology

If Marxism represents one of the most ambitious efforts to understand and advance the 'course' of history, it was its failure to understand and predict the conditions under which revolutionary action *failed to happen* that would lead to major efforts to develop alternative ways of knowing and changing the world. Nowhere was the paradigm put to a greater empirical test than during the 'Lost Revolution' of the German Weimar Republic between 1918 and 1923 (Morrow and Brown 1994). An issue that pre-occupied (and continues to pre-occupy) many Marxist theorists was the fact that the existence of class oppression in what had become a highly differentiated capitalist society had failed to yield the kind of revolutionary action that Marxism would have predicted and that (correspondingly) would have prevented the rise of Hitler and with it the isolation of the Soviet Union and (arguably) the rise of Stalin (Morrow and Brown 1994; Harman 1982).

Writing in long retrospect, Chris Harman (1982) suggests that the lack of revolutionary action reflected a more pervasive lack of coordination and leadership within the German Communist Party, which, he argues, could have 'produced a layer of militants' capable of toppling the Weimar regime. However, others began to reflect more critically about the cultural and psychological factors that affect the ways in which people experience exploitation and oppression in capitalist society. One early and important figure within this literature was Georg Lukacs (1885–1971), a Hungarian social theorist who shared with Hegel a desire to understand the broader ideological factors that shaped human agency and social identity in the context of capitalist social relations (Morrow and Brown 1994). For Lukacs, the idea that exploitation within capitalism thrives on the basis of 'external coercion' (or leadership) failed to capture what he felt were the dynamic ways in which individuals perceive and support the legitimacy of prevailing social orders. Reflecting upon the inability of the German working class to resist the rise of fascism in Weimar Germany, Lukacs argued that legitimacy and social order entailed a process of '*reification*,' whereby 'social agents came to identify falsely with a social reality that they perceived as "natural"' (Morrow and Brown 1994: 95).

Box 3.2 The lost revolution

aimed 'to make explicit the truth of primary experience of the social world' (Bourdieu 1977: 3), hermeneutics sought to recover the 'hidden truths and meanings from our everyday practices or from those of another age or culture' (Dreyfus and Rabinow 1983: xxiii).[43]

Influenced by the subsequent discovery of Marx and Engel's earlier and hitherto unpublished work on 'alienated labour,' Marxist scholars in the 1920s and 1930s also began to challenge the idea that Marxist theory and analysis could identify universal and predictive laws about the nature of history and society, and embraced instead a more critical or 'hermeneutic' interpretation of the way in which the lived experience of alienation may shape the dialectical relationship between individuals and society (Morrow and Brown 1994: Chapter 4).

Perhaps the most important manifestation of this was the so-called 'Frankfurt School' of Marxist scholars formed originally in 1923, and later forced into exile after the rise of Hitler (Morrow and Brown 1994). What made the Frankfurt School's work so important was its financial and therefore political autonomy from the German Communist Party and a (related) desire on the part of its principal researchers – Max Horkheimer (1895–1973), Theodor Adorno (1903–1969) and Herbert Marcuse (1898–1979) – to develop an empiricist approach to the study of class relations and revolutionary action (Morrow and Brown 1994). Central to their work was an effort to establish, through empirical research methods, the factors explaining the lack of revolutionary action between 1918 and 1923 (Morrow and Brown 1994).

With the release of Marx's unpublished work on alienation, with the 'lost revolution' in Germany and with the establishment of the Frankfurt School, Marxism began to move in directions that rejected the idea of establishing universal laws of history, and embraced instead a more critical interpretation of the ways in which cultural and psychological factors may affect revolutionary action.

As Morrow and Brown (1994) have argued, the Frankfurt School would have a lasting effect on two major bodies of Western European thought. One was the 'critical theory' of Jurgen Habermas and Anthony Giddens, whose work we take up shortly. A second was the post-structuralism of Jacques Derrida and Michel Foucault, whose impact we take up next.

The postmodern turn

Historically, post-structuralism emerged in critical response to the aims and assumptions of structuralism, a field that after the end of the Second World War began to dominate European (and especially French) social theory (see Box 3.3). Rejecting the idea that social action is somehow structured and constrained by universal codes of human discourse, writers like Derrida (1966) and Barthes (1977) began to look at the ways in which written texts and spoken discourse give rationality and meaning to social experience (Morrow and Brown 1994). Derrida (1966) for instance, questioned the idea of 'the self,' and argued that individuals embody multiple and often contradictory identities which affect the ways in which they interpret and act within the social world (Sarup 1989).

Influenced by Saussure, Derrida made an important distinction between the 'signifiers' we use to convey meaning and purpose and the images and ideas (i.e. the 'signified') we actually infer from social discourse. For Derrida, language was inherently 'unstable' in the sense that it conveys meaning (i.e. we understand the world through language), but through repetition, misinterpretation,

Although it is often associated with Marxism, 'structuralism' entails a much larger body of theory that aims to understand the basic elements that govern or 'structure' human behaviour (Morrow and Brown 1994). As Dreyfus and Rabinow have argued, structuralism entails an effort to 'treat human rationality scientifically' by searching for 'basic elements (concepts, actions, classes of words) and the rules by which they are combined' (Dreyfus and Rabinow 1983: xix–xx). Where phenomenology and hermeneutics are principally concerned with the ways in which individuals interpret social realities (and their practices within these realities), structuralism emphasizes the idea that a general theory of human action may be developed on the basis of linguistic and cultural interaction. One early and influential body of structuralism was linguistic structuralism, which held that scientific principles could be used to understand the grammatical rules and codes that structure the culture of human discourse. An earlier pioneer in the field was Ferdinand de Saussure (1857–1913), who argued that language could be divided between the 'surface features of speech,' what he termed *parole*, and the structural/grammatical rules of language systems (*langue*), which make language and communication possible (Morrow and Brown 1994).

Box 3.3 Structuralism

miscommunication *and writing*, it also constructs particular types of meaning that uphold certain views of the world, and exclude others. Language and (especially) writing, he argued, were 'intrinsically violent' (Parfitt 2002: 95) in the sense that they defined and classified what was reality.[44]

To counter the 'epistemic violence' of language and of knowledge, Derrida favoured a 'principle of least violence,' – an ethical position that would, in theory, liberate the conceptual and linguistic terms on which norms and ideas are communicated, and maximize the ability to recognize competing claims to truth and reality. Central to this philosophical approach was his theory of 'deconstruction,'

> … a method of reading the text so closely that the author's conceptual distinctions on which the text relies are shown to fail on account of the inconsistent and paradoxical use made of these very concepts within the text as a whole. In other words, the text is seen to fail by its own criteria.
>
> (Sarup 1989: 37)

By exposing the inconsistencies and contradictions of words, ideas and concepts, deconstruction would therefore yield a more democratic discourse about language and meaning (Parfitt 2002). However, the notion that a 'dialectic' of epistemic violence could be 'resolved' (in the Hegelian sense) was for Derrida a teleology in its own right, and therefore highly unlikely. On the contrary, Derrida argued, words and language always entailed an element of violence, and exclusion, which could be exposed (through deconstruction) but never eliminated (Derrida, cited in Parfitt, 2002). Knowledge, language and discourse therefore involved continual processes of deconstruction and reconstruction.

Foucault shared with Derrida a concern for the historical and institutional mechanisms by which discourses (of Enlightenment, madness, criminality, etc.) 'descend' and shape 'modern' social practice. For Foucault, discourses have a 'judicative' and a 'veridicative' function, suggesting the existence of terms and concepts that establish rules and norms concerning what may be included in a particular discourse (judicative) and what constitutes truth and falsehood (Parfitt 2002). In *The Order of Things*, Foucault argues that any claim to knowledge may be founded upon a series of 'serious speech acts,' statements which contain within them implicit and explicit norms upon which validity may be established by 'a community of experts' (Peet and Hartwick 1999: 130).[45] For Foucault, the transition to modernity entails a crucial intellectual and cultural break whereby the construction of knowledge goes from a 'natural' process of naming and ordering the world to an *episteme* that is structured by a set of highly specific rules and discourses which determine what constitutes valid ways of knowing.

By exposing the cultural means by which the 'human sciences' ascribe meaning and validity to symbols and language, Foucault's treatment of power, truth and knowledge therefore draws our attention to the social construction of knowledge and to the difficulty of using the language and methodology of modern science to understand the human condition. It also brings to light the notion that knowledge constitutes a form of power, whose disciplinary discourses shape and define what is knowable and what is desirable within the *episteme*.

Postmodern politics: class, social consciousness and (class) struggle

Alongside these intellectual discussions and debates were a number of historical processes and events that, during the 1960s and 1970s, made the 'postmodern message' increasingly attractive to scholars and social activists jaded by the inability of Marxism to advance a clear and coherent alternative to modern capitalism. One was the perceived 'corruption' of state socialism. By the end of the 1960s, the notion that the Soviet Union and/or the People's Republic of China could advance a model of progress had been shattered by the revelation that these and other models of 'actually existing socialism' had allowed, and in many cases conspired to commit, unspeakable acts of human suffering, including *inter alia*, the Stalinist purges of the 1930s and 1940s, forced collectivization in China and the Soviet Union and the 'Great Leap' famine of the late 1950s (Bernstein 2005; Sarup 1989). Whether these atrocities reflected the limitations of 'actually existing socialism' or whether they reflected the historical paths taken (or not taken) during pivotal historical junctures (Harman 1982), the idea that socialism could advance the human condition on the basis of Marxism had, by the end of the 1960s, become increasingly untenable.

Reinforcing these doubts were the French student riots of 1968. As thousands of students converged on the streets of Paris (originally in response to student protests about the administration of courses and curriculum at the Universities of Paris and Nanterre), the French Communist Party and French trade unions advised their members to abstain from joining or formally supporting the students, substantially undermining the idea that orthodox Marxism and/or Marxist political parties could (or should) represent the interests and ideals of myriad social movements (Sarup 1989).[46] From the Party's perspective, the perception was that the aims and actions of the French students were chaotic, contradictory to the interests of organized labour and 'out of control.' However, for many others the decision to abandon the students gave the impression of a movement that had lost touch with the 'real' needs and interests of society, and that future efforts to advance a progressive social agenda would therefore need to take place outside of the traditional Marxist paradigm.

Within this context, post-structuralism and postmodernism offered an attractive challenge to the perceived determinism of the grand development orthodoxies, especially Marxism.[47] First, they questioned the ability of intellectuals and other revolutionaries to speak and act on behalf of the proletariat, the subaltern, the peasantry, etc. In *Hegemony and Socialist Strategy*, Ernesto Laclau and Chantal Mouffe argued that 'there is not *one* discourse and *one* system of categories through which the 'real' might speak without mediations' (Laclau and Mouffe 1985 [2001: 3]). In so doing, they challenged the idea that revolution and revolutionary consciousness of the working class would 'arise spontaneously' as a result of the underlying contradictions of capitalism or – more controversially – as a result of intellectual efforts to 'read in the working class its objective destiny,' (Laclau and Mouffe 1985 [2001: 85]). The 'direction of the workers' struggle,' they argued, 'depends … upon its forms of articulation within a given hegemonic context' (Laclau and Mouffe 1985 [2001: 87]).

Second, they questioned the idea that material forces and relations can exist independently of the concepts, ideas and discourses we use to describe them. A central assertion being made here was that Marxian concepts like class, modes of production, alienation, contradiction, etc., were essentially social constructions of theorists attempting to understand the complexity of reality. For instance, Barry Hindess and Paul Hirst (1977) questioned the foundational belief – of Marx and of orthodox Marxism – that material 'activity and intercourse' can ever fully explain the diversity, complexity and inter-subjectivity of human consciousness, arguing controversially (see below) that theories 'only exist' as discourses.

Third, scholars began to reject the centrality/totality of a narrative/strategy aimed at understanding/acting upon the contradictions/injustices of modern capitalism. A subject of considerable criticism was the idea that unseen, historical forces and contradictions may evolve in ways that approximate the *telos* of orthodox Marxism and/or that these factors and processes may be associated with the 'needs' of the capitalist system or 'totality' with historical events and agency. Deleuze and Guattari (1977), for instance, argued that Marxism was essentially a narrative of salvation, which transformed what was effectively an interpretive system into 'an instrument of political and physical domination' (Sarup 1989: 101). Along similar lines, Hindess and Hirst argued that Marxism advanced 'an extremely limited range of economic class relations and the consequent neglect of the conceptualising of more complex forms of class relations' (Hindess and Hirst 1977: 2).

In short, the postmodernism of Derrida and Foucault gave rise to a deeper mood of skepticism which, in the context of 1968, began to challenge the idea that revolutionaries could speak on behalf of the proletariat, the peasantry, etc., and that class analysis may be used to reveal the underlying laws of capitalist development and social change.

The development 'impasse'

Within this dynamic social context, development sociologists began to question the theoretical relevance of using Marxism and dependency theory to understand and address the political economy of development. Drawing upon Laclau and Mouffe (1985 [2001]) and Hindess and Hirst (1977), David Booth (1985), for instance, argued that Marxist theories of dependency and development had become increasingly detached from the real needs and concerns of the developing world, displacing what he suggested were more fruitful lines of inquiry. Booth's central claim was that Marxian debates about capitalism and dependency had led to a series of intellectual 'cul-de-sacs,' stemming from what he perceived to be a 'meta-theoretical commitment' on the part of Marxist scholars to demonstrate the necessity of capitalist transition.

At the heart of this 'impasse' was the notion that Marxist and neo-Marxist (i.e. dependency) approaches entailed an ontology that failed to capture the complexity of (non-European) development and an epistemology that required an implicit or explicit comparison with an idealized and primarily European development experience. For Booth (1985), scholars were effectively 'reading off the text' of Marxist political economy, searching for instances of class formation and capital accumulation, without considering in sufficient detail the ways in which non-European factors (such as gender, religion and caste) may affect the nature and form of capitalist transition. Not only were they teleological, Booth argued, but the deterministic Marxism and structuralism of the 1960s and 1970s had 'been pursued at the expense of other, logically unproblematic and empirically challenging, research strategies' (Booth 1985: 772).

For Booth (1985), dependency scholars failed to define development or dependency in a way that could be validated (or falsified), and the inability to distinguish dependency from underdevelopment produced a circular reasoning in which underdevelopment was both the result of and a feature of unequal and dependent relations between the periphery and the core. Booth (1985) also highlighted the arbitrary nature of dependency concepts such as autonomy, exploitation and unequal exchange, which he argued were often characterized in relation to 'an ideal type of purely capitalist exchange' (Booth 1985: 771). Such comparisons, he argued, were developed on the basis of an arbitrary and often implicit assumption that Third World social relations could and should be understood on the basis of an idealized and de-contextualized understanding of capitalist social relations.

Booth's critique questioned not only the means by which Marxist and dependency scholars *classified* the units of their analysis, but also the ways in which they theorized the meaning and significance of historical processes and events. In so doing, he suggested that the tendency to privilege one theory or concept over another reflected an attachment (on the part of Marxists and of scholars in general it would seem) to a mode of explanation rooted in the assumption that societies are structured by rules and processes, which constitute a system, and that patterns and properties may be identified in relation to this system. Because they were so necessarily bound to the assumption that the spread of capitalist exchange relations would *or should* lead to the spread of capitalist relations of production, Booth argued, scholars became disproportionately obsessed with the seemingly unusual and 'continued' existence of peasant enterprise and informal non-capitalist relations.

Subject to criticism was the idea that the most deterministic variants of Marxist political economy had embraced a worldview which conflated the 'needs' of the capitalist system with historical events and agency. Framed in this way, historical events and processes, such as the formation of a working class, the (related) disenfranchisement of the peasantry, the class struggle, etc., were all part of a larger narrative that explained the nature and form of capitalist transition. In other words, these processes *had to happen* because they were necessary for the development of capitalism (Wolf 1982 [1997]; Booth 1985).

Beyond the impasse: the end of ideology?

In retrospect, Booth's article (1985) helped to capture a more general feeling of skepticism about the idea of documenting or explaining social relations in relation to the imputed 'needs' of a social system or to a particular *telos* or end. However, the idea that Marxism was primarily or necessarily pre-disposed towards the kinds of systems-based teleology he describes in his essay was deeply problematic for a number of scholars. For instance, Stuart Corbridge (1990) argued that Booth ignored the traditions within Marxism that aim to understand the ways in which perceptions of alienation and agency may shape the dialectical relationship between individuals and society. Similarly, Richard Peet (1991) takes issue with Booth (1985) (and also with Corbridge's earlier work, *Capitalist World Development*, 1986) for his use of post-structural theory and, especially, of Hindess and Hirst (1977).

At the heart of Peet's critique (1991) is the idea that theories and concepts can exist 'outside' of reality, and can be assessed, validated and employed with no explicit connection to history or to historical modes of production. For Peet (1991), the work of Booth (1985) and especially Hindess and Hirst (1977) implied an artificial understanding of knowledge and epistemology, in which theories and concepts are somehow separate and distinct from the 'real world' of material production relations among human beings and between humans and nature.

Contrasting the work of classical Marxism with the structural-functionalism of Talcott Parsons, Peet makes the case that Marxism in fact provides 'a successful synthesis of structural imperative and human behaviour' (Peet 1991: 176), suggesting that Marxian understandings of alienation, exploitation and human nature had 'systemic' qualities and that systems do in fact have needs. These needs are reflective of systemic contradictions between the structural imperatives of capitalism and the struggle to change (through revolutionary action) or maintain the very conditions of life (Peet 1991). In his own words,

> ... this struggle is deeply felt exactly because it occurs in fundamental, essential activities – people's acts of collective self-creation. If multiple social contradictions become fused in common experience, revolutionary acts take place and society may change.
>
> (Peet 1991: 176)

Drawing upon the work of Eric Wolf (1982 [1997]), Peet argued that Marxism and a 'fine sense of historical and geographical detail' (Peet 1994: 340) could be combined to produce an analysis that could be sensitive to matters of context and contingency.

Others (such as Vandergeest and Buttel 1988 and Corbridge 1990) advanced the idea that the way forward entailed a stronger emphasis on the ways in which people actually experienced processes of exploitation, alienation and social change. Central to this perspective was the idea that class formations and social relations are shaped not only by historical social forces and processes but also by the efforts of individuals to understand and act upon the structures and processes that have meaning and bearing in their lives. One important source of inspiration in this regard came from Pierre Bourdieu whose conceptualization of *habitus* suggested that social structures are reflective of the durable routines and practices that satisfy through social means the material needs 'characteristic of a class condition' (Bourdieu 1977: 72). A similar argument was advanced by Anthony Giddens (1979, 1984), whose theory of 'structuration' suggested that social structures are shaped not only by the forces and contradictions of capitalism, but also by the conscious effort of individuals to understand and act upon these structures. 'Structuration,' Giddens (1979: 5) argues, is rooted in the (seemingly universal) premise that every actor 'knows a great deal about the conditions of reproduction of the society of which he or she is a member.'[48]

Rejecting the idea of reducing social experiences and realities to the imputed needs of systems and forces, scholars also began to advance the idea of defining social concepts and histories in relation to a specific and primarily local social context. For instance, Vandergeest and Buttel (1988) argued that conceptualization and explanation of social phenomena may be deeply grounded in a reflexive phenomenology of lived experience:

> ... development sociologists must take into account why people do as they do in terms of the subjective meaning attached to what they do, rather than

simply explain all action by appealing to economic or other formal laws and models of society,

(Vandergeest and Buttel 1988: 690)

To pursue these lines, the authors made the case that 'power in particular cases (could be) conceptualized in terms of how class, gender, ethnicity and other sectional relations are experienced by the historical actors themselves' (Vandergeest and Buttel, 1988: 688), suggesting that structure still *mattered* to the explanation, even if structure was defined, interpreted and negotiated in conjunction with local actors and contexts. Similarly, Booth (1985) argued that social relations and conditions could be understood primarily in terms of their specific social context, and not in relation to what he suggested was a foreign or idealized understanding of development and change.

By rejecting the more functionalist and deterministic forms of Marxist teleology, scholars embraced a new form of 'post-Marxism,' in which the concepts of Marxism (i.e. surplus value, labour theory of value, etc.) could be used to understand (but not necessarily advance or address) the dialectics of capitalism. Stuart Corbridge helps to capture the essence of this important idea:

> Post-Marxism suggests both an intellectual tolerance of non-Marxism and a recognition ... that Marxism functions best as a critique of capitalism and not as a blueprint for socialism. More exactly, post-Marxism accepts that regimes of accumulation under capital are contradictory, are founded on asymmetry and are prone to crisis; but it denies that these contradictions work to a consistent set of spatial outcomes or lead to some form of apocalyptic denouement. Post-Marxism encourages us to see differences as well as uniformities; it returns to radical development studies a sense of time and a sense of place.
>
> (Corbridge 1990: 634)

The literature that emerged in critical response to Booth's original essay therefore challenged the idea that Marxism was only or necessarily pre-disposed towards determinism and teleology. For many scholars, the way forward entailed a deeper emphasis on the ways in which people actually experienced processes of exploitation, alienation and social change. However, the idea that Marxian concepts of labour, capital and class could be employed without an explicit ideological commitment to class struggle and revolutionary change was for a number of scholars deeply problematic.

The 'people without history': weapons of the weak or a weak weapon?

The retreat from grand social theory entailed a strong desire among many scholars to document what Eric Wolf (1982 [1997]) has called the 'people without history.' For instance, Charles Tilly argued in 1985 that the tasks of social history were (1) to document large structural transformations, such as the development of capitalism and the formation of nation states; (2) to reconstruct 'the experiences of ordinary people in the course of those changes'; and (3) to connect the two (Tilly 1985: 31; cited in Skocpol, 1987: 22). Similarly, Wolf's history of 'the people without history' (Wolf 1982 [1997]) starts from the premise that scholarly treatments of non-European history tend to extrapolate the events and processes (and analytical concepts) of the Europe experience(s) to the periphery of world economic power and scholarship:

> ... when social scientists began to examine these new men more closely, they treated them mainly as social problems – problems created by a severance from their roots through detribalization or immigration – rather than as social actors in their own right, responding to new conditions ... Research thus concerned itself mainly with what was absent – conditions and characteristics that had once been and no longer were, or conditions yet to come. Less was said of what was present, the relational matrix and context of working class existence.
>
> (Wolf 1997: 354–5)

Influenced directly by Wolf (1969, 1982 [1997]) and Tilly (1984), as well as Giddens (1979) and Bourdieu (1977), James Scott (1985) aimed to document the lived experiences of those 'lost' in the grand histories and meta-narratives of social science theorizing about socio-economic development and change. Particularly glaring for Scott (1985: xv) was the tendency among scholars 'on the left' to glorify large-scale peasant rebellions, and to assign to them a relevance defined only in relation to wider epistemic and geo-political concerns about security and/or national liberation. Of central importance to Scott was the idea that large peasant rebellions – of the kinds he considered in *The Moral Economy of the Peasant* (see Box 3.4) – were rare, and they rarely reflected the 'real' interests of peasants. Far more common, he argued, were the 'everyday forms' of resistance, such as 'foot dragging,' 'false compliance, pilfering, feigned ignorance' and so on (Scott 1985: xvi).

Scott therefore wanted to document and explain 'the prosaic but constant struggle between the peasantry and those who seek to extract labour, food, taxes, rents, and interest from them,' (Scott 1985: xvi). Towards this end he spent two years in a village in the Malaysian state of Kedah. The selection of the village enabled Scott to document, through informal interviews and participant observation, the subtle ways in which innovation (through double cropping) and mechanization (with the introduction of combine harvesters) of agriculture affected class relations, class conflict and associated practices of domination and resistance. In his own words:

> The struggle between rich and poor in Sedaka is not merely a struggle over work, property rights, grain, and cash. It is also a struggle over the appropriation of symbols, a struggle over how the past and present shall be understood and labeled, a struggle to identify causes and assess blame, a contentious effort to give partisan meaning to local history.
>
> (Scott 1985: xvii)

Drawing directly upon the work of Giddens (1979) and Bourdieu (1977), Scott aimed to move beyond the kinds of ideological/theoretical construction 'in which actors conform to a normative order that is somehow outside and above themselves' (Scott 1985: 305). Here he was referring partly to the structural functionalism of Talcott Parsons, but also to the Marxian idea that the peasantry constitutes a class whose false-consciousness, mystification and identity impede socially progressive forms of action. Like Giddens (1979) and Willis (1977), he assumed that social relations of domination and resistance are embedded in and contingent upon the specific ways in which local actors construct, interpret and respond to 'commonplace material practices':

> The main point for my purposes is that the peasants of Sedaka do not simply react to objective conditions *per se* but rather to the interpretation they place on those conditions as mediated by values embedded in concrete practices.
>
> (Scott, 1985: 305)

Scott's analysis advanced an ontology that left open theoretical and ideological assumptions regarding the nature of class and the historical conditions under which the development of particular forms of capitalist relations of production and exchange would take form. However, the emphasis on local, specific struggles to define social realities appeared to undermine the ability of scholars, researchers, activists, etc., to extrapolate insights from local and locally defined research that would, in theory, inform future courses of action.

Perhaps the most critical response to the kinds of ethnography advanced by Wolf, and especially by Scott, was Tom Brass (1991), who argued that Scott had shifted the intellectual and ideological focus away from the large and theoretically ambitious terrain of socialist revolution to what he felt was more mundane politics of local resistance, foot-dragging and survival:

> Unlike Lenin, for whom capitalist development not only benefits rich and poor peasants in different ways but also prefigures socialism, for 'moral economists' it merely provokes a return to pre-capitalist socio-economic structure, and thus cannot prefigure anything.
>
> (Brass 1991: 182)

For Brass (1991: 179), the political economy advanced by Scott (1985) and Wolf (1969) understated dramatically the ontological connection between capitalist modes of production and the historical formation of class relations and identity. Associating *Weapons of the Weak* with Scott's earlier work (1976) on peasant rebellions in Southeast Asia (see Box 3.4), Brass argues that Scott's conceptualization of class reified an ahistoric and 'natural' ethic of subsistence, in which the peasantry is somehow transformed from 'passive accepters of existing ideology into its active challengers' (Brass 1991: 182). Brass associates the work of Wolf (1971) and Scott (1985) with Chayanov's idea of an independent middle peasant, which is 'not located between the rich and the poor peasantry but

As the United States became increasingly engaged in Vietnam, American sociologists, anthropologists and political scientists became increasingly interested in the causes and nature of rural insurgency, especially in Southeast Asia (Salemink 2003). Now widely used as a tool for teaching the social and political impact of colonialism (often at the graduate level), the *moral economy debate* involved a series of essays written largely in response to *The Moral Economy of the Peasant,* published by James Scott in 1976. Drawing upon E. P. Thompson's 'moral economy' of the English bread riots, Scott (1976) argued that peasant rebellions in Indochina were primarily the result of French colonial policy that during the nineteenth and early twentieth centuries violated basic norms of reciprocity and welfare which, he suggested, were characteristic of traditional peasant society. Scott's central premise was that peasants tend to minimize the risks of economic failure by investing in risk-minimizing institutions and activities. 'Typically,' he argues, 'the peasant cultivator seeks to avoid the failure that will ruin him rather than attempting a big, but risky, killing' (Scott 1976: 4). This tendency towards subsistence has a profound effect on the ways in which individuals implement and perceive 'the technical, social and moral arrangements of a pre-capitalist agrarian order' (Scott 1976: 5). First, it creates incentives to implement and maintain institutions that reduce the risks of economic misfortune. Second, it creates a moral expectation that patrons will assist their clients during times of hardship and need (Scott 1976). The failure to meet this moral obligation (a result of a new capitalist orientation that reduces the importance of traditional patron-client ties), Scott argues, was the driving force that moved otherwise 'stable' peasant societies into rebellion (Scott 1976).

Implicit in Scott's analysis was the notion that colonial institutions governing land, labour and capital undermined pre-existing or 'pre-capitalist' institutions governing loyalty, reciprocity and exchange. This is believed to work in at least two ways. First, access to a new and externally sanctioned currency (in this case, state-sanctioned money) provides a means of obtaining important social resources, such as standing, loyalty and respect, without *necessarily* having to 'invest' in traditional and time-consuming rituals and rites of passage. Second, it creates an incentive structure in which economic relations were increasingly predicated upon the ability to accumulate, control and exchange 'commodities,' such as land, labour and capital (cf. Polanyi 1957: Chapter 6). Framed in this way, 'the development of land markets is assumed to be a disaster to the peasantry and is viewed as a product of outside social forces tearing asunder common property' (Popkin 1979: 69).

Drawing upon neo-classical theories of collective action and rational choice (e.g. Olson 1965), Samuel Popkin (1979) rejected Scott's essentialization of peasant society, and embraced instead the idea that the likelihood of peasant insurgency will reflect the relative costs and benefits of engaging in agrarian struggle. Framed in this way, rational decisions about collective action are governed by the incentives individuals (as opposed to classes or societies) have to support or resist the existing agrarian order. Popkin's main contention was that the willingness and ability to resist the penetration of markets in land would depend on:

- security of land title: 'when there is private land with secure title, the opposition to outside land ownership becomes far less pronounced' (Popkin,1979: 65);
- the ability to understand and penetrate key land-granting institutions within the state;
- the resultant demand for labour;
- the ability to access and control key marketing arrangements, all of which determine;
- the ultimate distribution of benefits.

Popkin's findings were controversial – not least because they were funded by the American military (Salemink 2003). Set in a context that was becoming increasingly hostile towards America's involvement in Vietnam, they also suggested that potentially rebellious peasants could be pacified through market inducements, especially ones provided through secure land title. In so doing, they helped to legitimize the American government's effort to counteract rural insurgency through redistributive land reform (Putzel 1992; Salemink 2003). By contrast, Scott's analysis (1976) suggested that rural insurgency reflected not only the individual costs and gains of engaging in collective action but also a political effort to resist the violation of what he suggested were basic norms of fairness, reciprocity and social welfare. For this reason, his position was more strongly linked with the 'dissident view' (Salemink 2003: 186) that political resistance to French and American power in Indochina was legitimate.

For an excellent treatment of the historical, political and ideological factors that framed the 'Scott-Popkin' debate, see Salemink (2003).

Box 3.4 The 'moral economy debate'

much rather corresponds to a different sector in the rural economy, composed of independent smallholders who own their own land which they cultivate with family labour' (Brass 1991: 175). For Brass, the intellectual origins of Scott's misrepresentation of Marxist political economy are rooted in the failure of orthodox Marxism 'to develop an adequate theory of class-determined ideological forms and practice,' and in the related impact of 'non-Marxist post-modernists who now use it to throw doubt on the current and historical existence of class itself' (Brass 1991: 180).

Instead of rural mobilization, Brass concludes, Scott's analysis draws our attention to the 'everyday' forms of resistance that 'prefigure' locally contingent forms of hegemony and domination. In so doing, he echoes wider concerns that the focus of Scott (and of many others) on interpretation and 'thick description' had, in effect, abandoned the ideological and theoretical commitment to the principles of Marxist political economy (cf. Leys 1996). As John Harriss (1994) has argued, he also takes a number of liberties with 'postmodernism,' and with Scott's analysis. First, the notion that Scott (1985) was only or primarily concerned with a 'middle peasant perspective' understates dramatically the very strong emphasis he puts in his analysis on the efforts of agricultural workers to interpret and resist the introduction of herbicides, combine harvesters and other forms of labour-displacing technology (cf. Harriss 1994). Second, to say that Scott (1985) had abandoned 'the current and historical existence of class itself' (Brass 1991: 180) suggests (falsely) that Scott's analysis had nothing to do with class. Upon closer inspection, it is clear that Brass' main problem with Scott's analysis concerns its inability to connect the struggles and processes he describes in his analysis with a wider ideological agenda, particularly one that would embrace Marxist and Leninist theories of social and revolutionary change.

Concluding remarks

Among the 'grand' theories of development, Marxism represents what is possibly the most ambitious effort in human history to understand and advance the 'course' of history. Rooted in the belief that human suffering results not from a timeless understanding of human nature, nor from a natural process of evolution, but rather from a particular history of class struggle, Marxism offered a powerful means by which human societies could liberate themselves from the exploitation of modern capitalism. However, the notion that history could be studied *and changed* through a science of history raised difficult questions about (1) the idea that the study of history may reveal underlying trends and universal laws on which future action may be based; (2) the extent to which human agency and free will may inform the construction of theory and practice; and (3) the extent to which individual subjectivities, ideologies and intellectual practices may affect the interpretation and construction of history.

By the end of the 1970s, the notion that Marxism could offer a viable means of understanding or addressing the political economy of development had lost considerable ideological appeal.[49] Alongside the interpretive trends that were now sweeping the social sciences, the 'lost decade' fostered new doubts about the ability of states and societies to create the conditions for material progress and social change. Influenced by the work of Anthony Giddens (1979, 1984) and Pierre Bourdieu (1977), scholars like David Booth (1985), James Scott (1985) and Eric Wolf (1982 [1997]) began to question the notion that class formation and social change could be modelled on the basis of an essentialized image 'in which actors conform to a normative order that is somehow outside and above themselves' (Scott 1985: 305). In so doing, they pioneered new ways of conceptualizing and documenting the ways in which people actually experienced processes of poverty, exploitation and social change.

In retrospect, the 1980s and 1990s were liberating times for the study of development (Booth 1993; Schuurman 1993; Harriss 2005). Reflecting upon his original thesis, David Booth argued in 1993 that 'the heavy atmosphere of intellectual stagnation' he described in 1985 had to a very large degree been replaced by an agenda that now studied and appreciated the complexity and diversity of the 'social, political and spatial' aspects of development (Booth 1993: 52). Empirical research on issues concerning race, gender and the rise of the Asian NICs, he argued, had produced new and important insights about the nature of capitalism, development and 'historically grounded variations in national economies' (Booth 1993: 53). Others have reached similar conclusions about the period in question: Frans Schuurman (1993), for instance, has argued that the empiricism of the 1980s and 1990s fostered new directions in development theory, including the regulation school, actor-network theory and the study of new social movements in Third World politics. Similarly, John Harriss has more recently called the period 'a time of creative research and the flourishing of new ideas' (Harriss 2005: 33), highlighting theoretical advances made on the basis of empirically grounded research on globalization, gender, famine and the environment.[50]

Within feminist theory, writers like Naila Kabeer (1994) began to highlight the ways in which gender inequalities affected the distribution of labour and income within the household. In so doing, they fostered new insights about the ways in which gender ascribes roles to men and women, and how the production and reproduction of gender roles may shape and constrain the range of economic opportunities for women, men and for society at large. Similarly, within sociology, anthropology and political science, scholars began to challenge the idea of reducing social movements to 'simple' social constructs, such as scarcity or class, highlighting instead the ethnic, regional and religious cleavages that underlie environmental and other 'new' social movements (see, for instance, Watts 1983; Peet and Watts 1996; Walker 2000; Vandergeest 2003; Agrawal 2005).

Underlying the retreat from the 'grand' theoretical narratives of Marxism and dependency was a fundamental desire on the part of scholars like Scott and Wolf to give the objects of social science (i.e. 'ordinary people') the opportunity to define through their own interpretations and practice the terms on which their values, experiences and histories were used for the purposes of social science scholarship. In so doing they sought to construct a theory that would account for action, revolution, etc., on terms that their respondents could recognize and understand.

However, the notion that history and hermeneutics could be combined to inform social theory appeared to imply an ideological retreat from the larger structural and historical transformations that were now shaping the political economy of development. Colin Leys (1996), for instance, questioned the idea that the grand total of 'mini-narratives' of local case histories would ever lead to the construction of new or 'better' theory. Particularly troubling for Leys was what he felt was a lack of explicit connection between the descriptive and explanatory detail of micro-level analysis and larger-scale processes of development and change.

Within this latter critique we can detect the writing of post-colonial authors such as Spivak (1988) and Said (1979), in which social science methodologies are challenged for extracting insights and knowledge of poor and marginal peoples, and using these 'data' to advance the careers of scholars and practitioners in primarily industrialized countries (Kapoor 2002, 2004). We can also detect a hostility towards the application of 'foreign' concepts and agendas to issues and events that are best understood on their own terms, which are assumed to be historically specific and contingent.

The challenge of forging a path between critical awareness and social action we shall take up in Chapter Four.

4 Development as discourse

Contesting the politics of
'post-development'

I can understand why words emptied of meaning serve the interests of the corporation and the state, but they don't 'enhance' or 'empower' people who would find in their freedoms of thought and expression a voice, and therefore a life, that they can recognize as their own.

Lewis H. Lapham (2006: 15)

The human word is the power that orders our chaos.

Northrop Frye (1958: 23)

It makes no difference whether I write or not. They will look for other meanings, even in my silence … So I might as well stay here, wait, and look at the hill. It's so beautiful.

Umberto Eco, *Foucault's Pendulum* (1989: 533)

Introduction

The previous chapter described the historical factors and debates that led to the meta-theoretical retreat from the grand Marxian narratives of economic development and social change. This chapter now considers a more critical body of scholarship, which shares with postmodernism a rejection of the modernist faith in the assumption that positivist social science may be used to advance the goals of universal progress and social change. Jaded by the experience of structural adjustment, environmental degradation, HIV/AIDS and the more general impression that coordinated efforts to address these and other kinds of suffering had failed to meet their explicit objectives, writers like Arturo Escobar (1995), Gustavo Esteva (1992) and Wolfgang Sachs (1992) began to suggest that development efforts (and the study of development efforts and outcomes) were part of a wider and primarily Western agenda whose discursive practices and interventions subject poor and politically marginal groups and nations to the interests, needs and ideologies of large development bureaucracies, such as the IMF, the United Nations and the World Bank. Central to this philosophical and political orientation was the notion that development can be discerned through the discourses that structure and give meaning to development policy and practice. As Jonathan Crush has argued,

> ... the primary focus is on the texts and words of development – on the ways that development is written, narrated and spoken; on the vocabularies deployed in development texts to construct the world as an unruly terrain requiring management and intervention ...
>
> (Crush 1995: 3)

Influenced by the work of Michel Foucault, post-development inspired a difficult assessment of the ways in which classical social theory could inform the study of development, fostering a deeper mood of ambivalence and skepticism about the (seemingly) humanitarian motivations of development agencies and projects. Cast in its most favourable light (e.g. elements of Peet and Hartwick 1999; Brigg 2002; Ziai 2004), post-development highlights the unstated aims and assumptions of development, which include both the 'old' development orthodoxies of modernization and central state planning (cf. Scott 1998), as well as the 'new' development orthodoxies of neo-classical economics and neo-liberal reform. By emphasizing the discourses that give meaning and power to development, it also draws our attention to the ways in which researchers legitimate their own work in relation to the norms and practices of academic sub-fields and disciplines, and to the moral terms on which academic researchers extract time, information and other forms of data from their informants in the field. In so doing, it helps to illuminate the historical and epistemic conditions under which 'development' emerged as a body of theory and practice.

However, the notion that development may be divined and realities captured through its discourses poses a number of philosophical and methodological problems. First, the idea that development (and reality) is constituted through discourse blurs the distinction between things whose value is real or (dare we say) universal (e.g. good health, happiness, freedom from physical pain) and ones whose value is socially constructed (cf. Lehmann 1997; Parfitt 2002; Nederveen Pieterse 1998, 2000). Second, and related to this last point, the notion that discourses are by their nature imbued with power and interest makes it difficult to differentiate 'good' social movements from ones that are not so good (cf. Parfitt 2002). Taken to the extreme, the notion that development is inherently biased (against the poor, against the 'South,' etc.) appears to foster a relativism that questions not only the dominant discourses and practices of the IMF, the World Bank, etc. but also called into question any effort to improve the human condition (Parfitt 2002; Nederveen Pieterse 1998, 2000; Corbridge 1998).

Our aims in this chapter are twofold. One is to understand and assess the major ways in which Foucauldian discourse analysis has been used to understand processes of economic development and social change. A second is to evaluate the political viability and desirability of using post-development to advance a critical and coherent agenda for social change.

The following section outlines Foucault's treatment of history, genealogy and power, examining critical interpretations from Marxist and liberal perspectives.

'The Foucault effect': history, genealogy and 'bio-power'

As noted in Chapter Three, Foucault's treatment of history, knowledge and power raises difficult questions about the nature of consciousness, agency and social action. By exploring the social transformations that would lead to the establishment of the clinic, the asylum, the prison, etc., Foucault aimed to isolate the historical breaks and discontinuities that account for modern discourses of power, classification and treatment. As Madan Sarup has argued, Foucault rejects the idea of a Hegelian teleology, 'in which one mode of production flows dialectically out of another' (Sarup 1989: 63–64), but employs instead a Nietzschean tactic of critique through the presentation of difference:

> Whereas traditional or 'total' history inserts events into grand explanatory systems and linear processes, celebrates great moments and individuals and seeks to document a point of origin, genealogical analysis attempts to establish and preserve the singularity of events, turns away from the spectacular in favour of the discredited, the neglected and a whole range of phenomena which have been denied a history.
>
> (Sarup 1989: 64)

In *Madness and Civilization* (Foucault 1961 [1965]), Foucault emphasizes the idea that the treatment of madness after the seventeenth century transforms from 'the relatively open and freewheeling world of the sixteenth century' (Sarup 1989: 105) to a culture rooted in the classification and clinical treatment of what becomes a social and medical 'problem.' Particularly important for Foucault is the 100-year period between the seventeenth and eighteenth centuries – a period during which the insane are confined first in the workhouses of seventeenth century Europe and then in the medical asylums that begin to appear during the eighteenth century. For Foucault, the pivotal break coincides with the establishment of a social and professional regime of observation, classification and treatment, whereby the 'patient' goes from being an outcast living on the fringes of European society to a subject of medical treatment and scrutiny. The individual is therefore not the living embodiment of reason or of the prevailing mode of production (as liberalism and Marxism would lead us to believe), but rather the product of a *dispositif* of 'subjectifying practices (discourses, institutions, architectural arrangements, regulations, laws, administrative measures, scientific statements, philosophic propositions, morality, philanthropy, etc.) which had been instrumental in forming the modern individual as both object and subject' (Dreyfus and Rabinow 1983: 120–1).

Central to Foucault's later rendering of the individual and of society is his concept of bio-power. Broadly speaking, bio-power describes the ability to decide not only the moral and technical terms on which life may be taken (through, for instance, execution, warfare, etc.), but also the moral and technical terms on which it may be sustained. As Achille Mbembe has argued, the power being described in this instance implies the power to divide people:

... into those who must live and those who must die. Operating on the basis of a split between the living and the dead, such a power defines itself in relation to a biological field – which it takes control of and vests itself in. This control presupposes the distribution of human species into groups, the subdivision of the population into subgroups, and the establishment of a biological caesura between the ones and the others.

(Mbembe 2003: 17)

Although bio-power may manifest itself in many different ways, its impact may be most readily discerned by exposing the instances during which individual freedom is most profoundly defined and constrained through modern technologies of observation and control. For Foucault, the most illustrating examples of bio-power were therefore those that conferred the greatest powers of ordering and observation, that is, the military, the factory and (most famously) the prison, where:

The exercise of discipline presupposes a mechanism that coerces by means of observation; an apparatus in which the techniques that make it possible to see induce effects of power, and in which, conversely, the means of coercion make those on whom they are applied clearly visible.

(Foucault 1975 [1977: 170–1])

The crucial point here is that the disciplinary technology of the nineteenth century did not arise in response to a pre-existing problem but, rather, it created categories of criminality and began to exercise through classification and observation new forms of social control (Parfitt 2002: 46). As Dreyfus and Rabinow argue, Foucault is most definitely not telling the story of scientific progress:

Rather, the story for Foucault is the other way round. It is in the first major moves toward social internment, toward the isolation and observation of whole categories of people, that the first glimmerings of our modern medical, psychiatric and human sciences are to be seen.

(Dreyfus and Rabinow 1983: 5)

However, bio-power is not only confined to the formal hierarchies of observation, bureaucracy and control. In later writing, Foucault (1978 [1990]) emphasizes the idea that coercion and repression also exist in the multiple 'force relations' that determine and define what constitutes 'acceptable' social conduct. In *The History of Sexuality*, for instance, Foucault (1978 [1990]) emphasizes the 'repressive hypothesis' that sexuality is regulated and subdued not (only) by modern bureaucracy but also by the family, by peers, by advertising, by religion and by a potentially infinite number of groups and relations whose cumulative effect is the definition and regulation of social norms governing sexual expression and behaviour. Framed in this way, power rests not in 'a general system of domination exerted by one group over another,' (Foucault 1978 [1990: 92]) but rather in a wide and diffuse range of social norms and relations that define, censor and treat what constitutes sexually deviant social behaviour.

Debating Foucault

Foucault's rendering of history, genealogy and power engendered substantial controversy and debate, especially within Marxism. First, his focus on the 'local, specific struggles' (Sarup 1989: 88) to define through discourse the nature of madness, criminality and sexuality implied for many Marxist scholars what appeared to be a trivial shift away from the basic focus of historical materialism on class and the dialectic of class struggle. Not only did it shift the analysis away from the economic realm of material social relations, it did so in a way that emphasized issues and actors (e.g. madness, criminality, sexuality) that were, from a Marxist perspective, tangential (at best) and trivial (at worst) to the analysis of class relations and social change. In *Madness and Civilization*, for instance, Foucault (1961 [1965]) describes *in great detail* the emergence of a new rationality whose practice would facilitate and explain the diagnosis and treatment of madness in the age of reason. Consider, for instance, the following quotation:

> The paradox of the hot bath is symmetrical: it draws the blood to the periphery, as well as the humors, perspiration, and all liquids, useful or harmful. Thus the vital centers are relieved; the heart now must function slowly; and the organism is thereby cooled.
>
> (Foucault 1961 [1965: 169–70])

Foucault's point here was to make the historical conditions explicit under which particular (i.e. 'modern') perceptions and treatments of madness began to emerge. However, for Marxists, the link between this kind of 'thick description' and the dialectic of class struggle suggested a nuance that was difficult if not impossible to decipher.

Second, and especially in his later work, Foucault's concept of bio-power renders problematic the idea that power is embodied in and applied by a single state or an over-arching ideology. As Sarup has argued, power for Foucault is 'much more ambiguous, since each individual has at his or her own disposal at least some power' (Sarup 1989: 87). Framed in this way, power serves not only the reproduction of the material relations of production, but also the subjectivities that structure the myriad social relations between individuals and society. As David Harvey has argued,

> The prison, the asylum, the hospital, the university, the school, the psychiatrist's office, are all examples of sites where a dispersed and piecemeal organization of power is built up independently of any systematic strategy of class domination. What happens at each site cannot be understood by appeal to some overarching general theory. Indeed the only irreducible in Foucault's scheme of things is the human body, for that is the 'site' at which all forms of repression are ultimately registered.
>
> (Harvey 1990: 45)

The implication is therefore that power is dispersed through other 'already existing power networks, such as the family, kinship, knowledge, and so forth' (Sarup 1989: 87), and that power is most effective and 'insidious' when it includes, intervenes and even assists.

Third, and related to this last point, Foucault's rendering of power and of history appears to undermine the idea that a collective realization of a social condition (i.e. exploitation) may constitute a political power (of class consciousness) that could in theory resist or challenge the power and injustice of capitalism. For Foucault, power inheres not only in the bureaucratic and administrative channels of private property, labour regulation, etc. but also in the 'hierarchies and structures of power' (Sarup 1989: 86), which underlie the revolutionary movements and parties that claim to represent the proletariat. The implication here is that the ability of vanguard parties and movements to speak 'on behalf' of the proletariat is itself structured by myriad subjectivities and power relations within society, rendering problematic the legitimacy of any social movement (Sarup 1989). For Spivak (1988), the interpretation advanced by Foucault misconstrues dramatically Marx's own rendering of the relationship between class position (an empirical fact) and class consciousness (the recognition of a shared condition of exploitation and identity):

> ... the development of a transformative class 'consciousness' from a descrip-
> tive class 'position' is not in Marx a task engaging the ground level of
> consciousness. Class consciousness remains with the feeling of community
> that belongs to national links and political organizations, not to that other
> feeling of community whose structural model is the family.
>
> (Spivak 1988: 277)

Fourth, and from a broader ideological perspective, Foucault's genealogy advances what appears to be a relativist understanding of social movements and social change (Giddens 1984; Habermas 1987; Sarup 1989; Parfitt 2002). Jurgan Habermas (1987), for instance, has argued that Foucault's analysis lacks a foundational normative basis upon which we can assess the nature of power. Lacking foundational statements that would distinguish between acceptable and unacceptable forms of power, Parfitt argues, 'we have no grounds for resisting oppression, or for determining which forms or movements are genuinely emancipatory and which are not' (Parfitt 2002: 51–2).

Fifth, and finally, it has been argued that Foucault lacks a theory of action, which renders problematic the production of discourse that creates the modern human subject (Parfitt 2002; Habermas 1987; Sarup 1989; Rossi 2004; Taylor 1991). Madan Sarup, for instance, has argued that Foucault's effort to 'rid ourselves of the constituting subject' (Sarup 1989: 89), gives the impression of a world in which power, society and societal change are entirely the result of discourse and 'strategy'. On one hand, Foucault suggests that human agency is essentially the product of modern rationality and bureaucracy, but on the other he implies an original state in which the freedom to act has been undermined by modern efforts to order and reform the individual. As Parfitt (2002) argues, the implication is that the condition being treated, structured and classified by discourse had an existence that was prior to the imposition of modern power and discourse.

Whether all of these criticisms may be levelled against *all of* Foucault's writing, they certainly help to illustrate the challenge of reconciling a post-structural critique with an agenda of human emancipation and social change.[51] In what follows, I shall argue that post-development has grappled with similar problems concerning the ontological nature of class, history and progress. On the one hand, it defines itself primarily in relation to the voices and experiences of those undervalued, unrecognized and *compartmentalized* within the dominant (and often Marxist) development discourse. On the other, it struggles to articulate with sufficient clarity or purpose the ways in which it would address the needs (either real or perceived) of those with whom it claims solidarity. The result is a field that either engages in simplified rhetoric about the underlying nature of capitalism, modernization and 'the West' or one that obviates itself of any apparent need to connect theory and praxis.

To make my case, I distinguish between two 'types' of post-development. One, exemplified in the work of Arturo Escobar (1995), traces the broad historical contours of development, and suggests that the discourses put in place by national governments and international development agencies are *in effect* ideological and historical constructions, whose classification, measurement and treatment of development problems such as hunger, malnutrition, illiteracy, etc., are at best inappropriate and at worst deeply destructive to the 'real' needs and experiences of those objectified by development discourse. A second, exemplified in *The Anti-Politics Machine*, by James Ferguson (1990 [1994]), suggests that development may be used to legitimate and expand bureaucratic power, but that these processes are not only or necessarily detrimental to the poor. Indeed, Ferguson's analysis suggests that the power of development may be most effective (and insidious) when it assists, and succeeds.

Like Escobar, Ferguson rejects the central claims and assumptions of orthodox Marxism, and favours instead an ontology that captures through discourse the constructed and contested nature of reality. However, unlike Escobar (1995), Ferguson makes no grand claims about the geo-political aims of development, about the imputed biases of development and about the associated need to develop an alternative to a dominant development discourse. On the contrary, he rejects the idea of connecting his analysis to an explicit political agenda, thereby avoiding the kinds of moral and political relativism that undermine the post-development project. Nevertheless, by distancing himself from the notion that scholarship should be engaged politically, Ferguson also 'depoliticizes' his own discourse, suggesting at the end of his analysis (in his 'Epilogue') that the best scholars can do to address the injustices they find in their research is to encourage a more critical awareness about the discursive power of development. In so doing, he illustrates the limitations of using post-structural theory to advance a critical and coherent vision of social change.

The chapter proceeds with the following section, which explores the theoretical aims and assumptions of post-development. Next, it reviews the literature that emerged in critical response to post-development, which I then compare with the work of Ferguson (1990 [1994]). Finally, I use more recent work by David Mosse (2005) and Stuart Corbridge *et al.* (2005) to illustrate the possibilities (and dangers) of using discourse analysis to advance a constructively critical perspective on development.

Development as discourse: the politics of post-development

Post-development may be usefully defined as a political and philosophical orientation that employs elements of discourse analysis to question the dominant aims and assumptions of development. As Trevor Parfitt (2002) has argued, its intellectual origins are rooted in the intellectual criticisms of post-structuralism, post-colonialism (see Box 4.1) and related fields of feminist theory and cultural studies which, during the 1960s and 1970s, began to take hold on university campuses and wider sites of cultural criticism in Western Europe and the United States.[52] Its political origins stem from the profound sense of disillusion that followed the monetary and fiscal crises of the 1970s and 1980s, broader trends which were happening in the social sciences and a related realization that existing theories of development and change – particularly ones deriving from Marxist political economy – were unable to account for the complexity and diversity of historical transformation, cultural variation and social change.

Like post-colonialism (see Box 4.1), post-development employs elements of discourse analysis to question Western narratives of imperialism, modernization and progress. However, in contrast to the post-colonial writing of Edward Said and Gayatri Spivak (see Box 4.1), it tends to be less concerned about the conditions and contested identities left in place by colonialism than it is about the 'colonizing' aspects of development. Historically, therefore, the focus tends to be on the period that begins after the end of the Second World War.

Like dependency theory, post-development *claims* strong political, cultural and epistemic ties with 'Southern' scholars and activists, whose rejection of the dominant neo-liberal orthodoxy and the intellectual roots of this orthodoxy engendered strong feelings of solidarity against the West. However, unlike dependency, which always embraced the idea of achieving improvements in material well-being, post-development also questions the ways in which 'problems' like poverty, vulnerability and illness are classified in relation to the perceived needs, goals, interests and practices of development projects and agencies and in relation to the perceived needs, goals and interests of those who would, in theory, receive the benefits of development.

By exposing and revising the core assumptions of development, post-development would produce a body of scholarship and practice that would recognize and respect the agency, identity and individual worldviews or narratives of those objectified by development. First, it would challenge an epistemology whose normative roots and motivations were hostile to an appreciation and authentic representation of culture, difference and diversity. Second, it would challenge the idea that the study and practice of development can or should be committed to the universalizing norms and discourses of a positivist social science, whose principal aims are to

Post-colonialism is a political and philosophical orientation that employs elements of discourse analysis to challenge the underlying myths and assumptions of European colonialism, history and identity. Seminal contributions in the field would include *The Wretched of the Earth*, by Frantz Fanon (1963) and *Orientalism*, by Edward Said (1979 [1994]), both of which argued that the power of European colonialism was rooted not only in the physical subjugation of conquered lands and peoples but also in the cultural production of a colonial discourse that denied the agency of subjugated peoples to define themselves and their futures in terms of their own. The power to rule was therefore most insidious when applied through the guise of benevolence, civilization and progress. Gayatri Spivak, for instance, uses Britain's efforts to abolish the *sati* (the practice of widow self-immolation) in colonial India to illustrate the ways in which acts of benevolence could be used to legitimate colonial power. By redefining formal codes of gender and the rights of widows, she argues, the abolition of the *sati* validated the authority of the British to govern and in Spivak's own words, to 'save' 'brown women from brown men' (Spivak 1988: 297).

Building upon the work of Fanon and Said, writers like Spivak, Ranajit Guha and Homi Bhabha aimed to re-constitute post-colonial histories by documenting 'subaltern' narratives about the colonial and post-colonial experience (Peet and Hartwick 1999). In the process, many questioned the normative principles upon which Western academics (or, rather, academics in Western universities) interrogate the non-western world for the purposes of academic scholarship. For instance, Spivak (1988) argues that any effort to let (or to make) the subaltern speak cannot possibly escape the norms and practices of ethnocentrism and inequality that shape the relationship between (primarily) northern researchers/ academics and the 'Third World Other.' By privileging the construction of theories whose norms and practices are defined primarily by academics and researchers in 'the core,' social science reinforces a Third World that facilitates the objectification of poverty, marginality, etc. and provides for the core the information, knowledge and data necessary for the construction of knowledge.

Sources: Spivak (1988); Peet and Hartwick (1999); Kapoor (2004)

Box 4.1 Post-colonialism

generate comparable and replicable propositions about the conditions under which societies may undergo development (and whose model, it was implied, was based primarily upon the North American and Western European experience). Finally, it would challenge the under-socialized ways in which development scholars and professionals defined and classified problems, conditions and people both in the context of development interventions and in academic research.

The following section considers the central claims and assumptions of post-development, first by reviewing its treatment of history and development, and then by assessing its agenda for social change.

Post-development histories

Post-development and postmodernism share the idea that history is partial, and necessarily biased in favour of Western views about modernization, Enlightenment and progress. Emulating Foucault's emphasis on historical breaks and discontinuities, writers like Wolfgang Sachs (1992), Gustavo Esteva (1992), Gilbert Rist (1997) and Arturo Escobar (1995) reject the idea that development may be understood as a benevolent response to the manifest needs of the world's poor, and suggest instead that 'development' was an idea that emerged in relation to the geo-political needs and interests of a world order that was beginning to form at the end of the Second World War. They all put strong historical and intellectual weight on US President Harry Truman's Point Four doctrine, which in 1949 outlined the Truman Administration's strategy for the post-war world, and articulated the prevailing view of poverty, inequality and progress in the emerging Cold War era. Central to this vision was the idea that: foreign aid could *and should* facilitate the liberation and development of non-European and *non-capitalist* societies; the United States could *and should* lead this process; and progress could *and should* be achieved through the application of reason and the enlightenment of *hitherto ignorant peoples and cultures*. Deeply embedded in the Point Four doctrine, they argued, were liberal understandings of freedom, Enlightenment and progress, romantic assumptions about the 'real' needs and aspirations of previously subjugated cultures and peoples and an effort to differentiate American strategic objectives from the end of European imperialism.

In *Encountering Development*, Arturo Escobar (1995) argues that development can be understood as a 'regime of representation,' whose discourses emerged during the historical processes of decolonization in Asia, Africa and Latin America, and the emergence of the United States as a global power.[53] 'Within the span of a few years,' he argues,

> ... an entirely new strategy for dealing with the problems of the poorer countries emerged and took definite shape. All that was important in the cultural, social, economic and political life of these countries – their population, the cultural character of their people, their processes of capital accumulation, their agriculture and trade, and so on – entered into this new strategy.
>
> (Escobar, 1995: 30)

The 'new strategy' entailed strong images of righteousness and responsibility, which redefined the West's understanding of itself and its relationship with the rest of the world. According to Escobar (1995: 31), 'These concepts did not exist before 1945.'

Starting from the premise that international efforts to promote development (e.g. the Marshall Plan, the United Nations, the Bretton Woods institutions) were never 'really' intended to help those ravaged by the violence and suffering of the Second World War and the Great Depression, post-development suggests that development was in fact a strategic response to the combined threat of communism, Third World nationalism and more general conditions of poverty and over-population in the post-colonial world. For Escobar (1995: 31–5), 'development' provided the ideological and instrumental means by which the United States and allied Western powers could legitimate and pursue what were in fact the real strategic aims of: (1) consolidating the 'core' capitalist economies; (2) establishing higher rates of profit for core capitalist interests; (3) securing strategic control over raw materials, especially oil; (4) expanding overseas markets; and (5) deploying 'a system of military tutelage' that would help to secure American interests abroad and generate wealth for its military industrial complex.

The development 'paradigm' was therefore one that upheld the needs and interests of American capitalism (especially at the end of the Second World War), Western values (defined very broadly) and an entire 'culture' of professional and academic ethics and practices, which were established and reified through the implementation of development projects and policy.

Normalization and discourse

Central to the institutionalization of American/Western power during the post-War era was a language (and a related body of practice that recognized and validated this language) that could justify through classification, measurement and intervention the application of what was in fact American/Western power. Tracing the history and impact of agricultural and nutritional programs in Latin America (primarily in Colombia), Escobar (1995: Chapter 4) argues that development interventions by the World Bank, by the Government of Colombia and by other relevant 'actors' had the 'instrument-effect' of 'maintaining certain relations of domination' (Escobar, 1995: 112), whose practices upheld both an economic paradigm that privileged the needs of domestic and multi-national capital over those of landless peasants and urban labour and an epistemology that excluded local and indigenous worldviews and ways of constructing knowledge:

> Only certain kinds of knowledge, those held by experts such as World Bank officials and developing country experts trained in the Western tradition, are considered suitable to the task of dealing with malnutrition and hunger, and all knowledge is geared to making the client knowable to development institutions.
>
> (Escobar 1995: 111)

At this dynamic historical juncture, Escobar (1995) argues, neo-classical economics quickly emerged as a dominant 'culture,' in which the central tenets of analytical Marxism (especially Marxian ideas of surplus value and the labour theory of value) were rejected in favour of a new emphasis on individual utility, economic rationality and market equilibrium. Remarkable in its own right, the rejection of Marxian theorizing was of less historical significance for Escobar (1995) than was the elevation of a 'culture' that aimed to achieve material improvements in human well-being through the application of theories whose pre-suppositions and causal forces were assumed to be universal.

Construed in this way, development renders certain categories of experience and concerns 'invisible' (Escobar 1995) to dominant development discourses. Drawing upon Foucault's genealogy of the prison (Foucault 1975 [1977]), Escobar (1995) suggests that development defines what is 'normal' and 'abnormal,' and in so doing devalues other forms of knowledge, values and experience. 'Peasants,' for instance, are seen in purely economic terms, as 'seeking a livelihood in the rural areas,' not as trying to make viable a whole way of life' (Escobar 1995: 162). Women's productive value is concealed, or distorted by the development process and by the development industry's efforts to understand and act upon the undervalued status of women in development (WID). Finally, the natural environment is undervalued – and therefore degraded – not necessarily as a result of institutional neglect, but more subtly, as a result of the development industry's efforts to manage and control environmental problems.

Starting from the premise that development is a political construction, post-development questions the idea that the conventional aims, measures and achievements of development (e.g. improvements in literacy and life expectancy, reductions in maternal and child mortality, etc.) are universally or even essentially desirable. For Escobar, the universalizing claims of neo-classical economics are particularly problematic in the sense that they deny 'the capacity of people to model their own behaviour and reproduce forms that contribute to the social and cultural domination effected through forms of representation' (Escobar 1995: 37). Although he concedes that the industry 'created new knowledge capabilities,' he argues that 'it also implied a further loss of autonomy and the blocking of different modes of knowing' (Escobar 1995: 37).

The new development paradigm therefore embraced an epistemology that aimed for generalization and the application of universal principles that were increasingly dependent upon neo-classical assumptions about individual utility, economic rationality and market equilibrium. Facilitating the spread of this development rationality was an expanding array of institutions whose practices were adopted, emulated and disseminated by the newly established international development bureaucracies.

Along similar lines, Gustavo Esteva and Madhu Suri Prakash (1998) suggest that the combination of international development and global capitalism constitutes a systemic or global force, whose impact on 'the South' has been universally detrimental to the 'real' needs and interests of those affected by such processes. In so doing, they make a qualitative distinction between '*social minorities*,' whose political and economic elite favour and support the 'global project' and the *'social majorities,'* the huddled masses of farmers, urban labourers, migrants, ethnic minorities, etc. whose needs and preferences are necessarily antithetical to the aims of 'the global project,' a term they use to characterize 'the current collection of policies and programs ... committed to the economic integration of the world and the market credo' (Esteva and Prakash 1998: 16). 'Local autonomy,' they conclude, is the 'only available antidote to the global project' (Esteva and Prakash 1998: 37).

Similar assertions can be found in *The Post-Development Reader*, edited by Majid Rahnema and Victoria Bawtree (1997). Incorporating an implausibly wide range of authors and perspectives,[54] the editors suggest that post-development constitutes a pervasive sense of malaise about the idea that development can or should aim to improve the lives of those less fortunate. To make their case, they marshal the work of Vandana Shiva (1997), Marshall Sahlins, Ivan Illich (1997), Rajni Kothari, Helena Norberg-Hodge (1997) and Gandhi (1997), among many others, to illustrate the ways in which efforts to achieve modernization and development have undermined the wisdom of 'vernacular society,' a term Majid Rahnema (1997a: 128) uses to describe the 'words and expressions' that are 'proper to the people using them naturally, as opposed to a language cultivated and borrowed from elsewhere.' Starting from the premise that 'development' is necessarily external and antithetical to 'vernacular society,' Rahnema then calls into question the entire history of planned development:

> The issue is, therefore, not that development strategies or projects could or should have been better planned or implemented. It is that development, as it has imposed itself on its 'target populations', was basically the wrong answer to their true needs and aspirations.
>
> (Rahnema 1997b: 379)

Leaving for now the (rather dubious) notion that people's true needs and aspirations may be represented in this way, a central point of departure for post-development is therefore the idea that the 'development' of national governments and international agencies undermines the cultures of societies previously untouched by the destructive forces of capitalism and modernity, and that an alternative to the mainstream of global capitalism and/or international development must be developed and pursued. But what form will these alternatives take? And how will they address the problems raised by post-development?

Visions of change

Explicit in the work of many post-development writers is the notion that efforts to address the problems created by development will entail a return to simple (and often rural) living idealized in the form of a pre- or non-capitalist existence. In *The Post-Development Reader*, for instance, Majid Rahnema (1997b: 399–402) suggests that efforts to right the wrongs of development will entail the establishment of an 'aesthetic order' that rejects the logic and rationality of development in favour of political and economic arrangements rooted in notions of harmony, sustainability and self-sufficiency:

> The post-development era is in dire need of a commitment from all good men and women to the creation of an aesthetic world order in which new forms of friendship and solidarity will be able to interact in order to stop the evil forces of the 'global village' destroying the last 'good people' struggling to protect themselves from them. It would be up to the emerging circles of friends to explore how the profoundly humane ethics of such an order could be reconciled with the unavoidable needs of a rational order.
>
> (Rahnema, 1997b: 400)

Along similar lines, Esteva and Prakash (1998) argue that solutions to the failures of capitalism and development will entail a 'downsizing' of social relations to what they call a 'human scale' of interaction.

Leaving for now the utopian ways in which politics and development are being construed in these instances, a related aim in the post-development literature is the idea that, by devising alternative ways of understanding and representing the world, local and indigenous social movements (and intellectuals aligned with these movements) would more faithfully capture the 'real' (and usually unstated) needs and interests of those excluded by dominant development discourses. In *Encountering Development*, for instance, Escobar suggests that development scholars need to recognize the 'plurality of models' that can be used to interpret and evaluate social and economic relations, and make explicit 'the mechanisms by which local cultural knowledge and economic forces are appropriated by larger forces' (Escobar 1995: 98). Echoing Spivak (1988), Escobar highlights 'a reluctance on the part of academic audiences in the First World ... to think about how they appropriate and "consume" Third World voices for their own needs' (Escobar 1995: 224). The solution, he argues, is to create the conditions under which alternative visions may be articulated in a way that represents the 'real' (and usually unstated) needs and interests of those excluded by dominant development discourses:

> One must then resist the desire to formulate alternatives at an abstract, macro level; one must also resist the idea that the articulation of alternatives will take place in intellectual and academic circles, without meaning by this that academic knowledge has no role in the politics of alternative thinking.
>
> (Escobar 1995: 222)

However, unlike Spivak (1988: 275), Escobar situates the unequal exchange of knowledge not in an 'international division of labour' but in a normalizing process of discourse and international intervention. For Escobar (1995), the solution to the problem of articulation and representation is to document and disseminate the ways in which 'marginal' groups resist the spread (or globalization) of development and market capitalism. The approach, he argues, will be inductive, and receptive to the specific needs and nuances of local context and history:

> The remaking of development must ... start by examining local constructions, to the extent that they are the life and history of a people, that is, the conditions of and for change.
>
> (Escobar 1995: 98)

Escobar's assertion is that any effort to reconstitute the study and practice of development must achieve what he suggests is the right combination of open inquiry and social resistance to 'global' capitalism. Echoing the kinds of post-structuralism we encountered in Chapter Three, he rejects the ontological assumptions of traditional 'political economy,' and favours instead a perspective that would 'deal with the cultural dynamics of the incorporation of local forms by a global system of economic and cultural production' (Escobar 1995: 98). 'A more adequate political economy,' he argues,

> ... must bring to the fore the mediations effected by local cultures on translocal forms of capital. Seen from the local perspective, this means investigating how external forces – capital and modernity, generally speaking – are processed, expressed, and refashioned by local communities.
>
> (Escobar 1995: 98)

Broadly conceived, post-development therefore renders problematic the idea that social science research may represent the 'real' needs and perceptions of those affected and disaffected by development. In so doing, it makes a powerful argument in favour of linking scholarship and praxis, ideally by incorporating poor and politically marginal voices into the construction of knowledge. However, the notion that scholars would incline towards a more open and/or 'local' resistance to the forces of global capital is of course a difficult question. The following section considers the challenges and contradictions that such an agenda would entail.

Debating post-development

Critical interpretations have challenged post-development for trivializing the achievements of development (e.g. Corbridge 1998; Harriss 2005); for blaming problems of poverty and inequality *solely* on the efforts, programmes and ideologies of (Western) development agencies (Corbridge 1998; Brigg 2002; Nederveen Pieterse 1998, 2000); for advancing a Romantic vision of human development (Nederveen Pieterse 1998; Corbridge 1998; Briggs and Sharp 2004); for grounding their analysis in what many felt is an ambiguous and relativist understanding of all that 'global capitalism' was not (Parfitt 2002; Nederveen Pieterse 1998; Corbridge 1998); and finally, for advancing an agenda that fails to articulate with sufficient clarity or purpose the ways in which post-development would or could move beyond the perceived failures of dominant development discourse.[55] Others (such as Sylvester 1999; Parfitt 2002; Nederveen Pieterse 1998; Goss 1996; and Briggs and Sharp 2004) questioned the notion (both explicit and implied) that post-development scholars were somehow better placed (geographically, culturally, ideologically) to speak on behalf of the poor.

One argument, which has been directed especially towards the work of Sachs (1992), Rahnema (1997a; 1997b) and Esteva and Prakash (1998), is that post-development offers simple and arbitrary distinctions between the development establishment (whose definition varies enormously) and those affected (often adversely) by development efforts. Reviewing primarily the work of Rist (1997), Esteva and Prakash (1998) and the edited volume by Rahnema and Bawtree (1997), Stuart Corbridge (1998) argues that post-development offers 'unhelpful and essentialised accounts of the West and the Rest,' and questions the notion that people in low-income regions and countries would naturally prefer the kinds of 'alternative' living being advanced by the post-development narrative. Along similar lines, Nederveen Pieterse (2000) argues that the Western/non-Western, foreign/local binary 'denies the agency' of the very societies and social movements post-development scholars are aiming to engage (cf. Peet and Hartwick 1999: Chapter 5).

The strong tendency of post-development to demonize development (and with it Western applications of modern science) simplifies the diversity and complexity that exists among development agencies (comparing the work of UNICEF, for instance, with that of the World Bank, or suggesting that all of the World Bank's activities can be categorized under a single narrative: Nederveen Pieterse 2000) and development scholars (as diverse in opinion and outlook as Rostow and Wade in Corbridge 1998). As Parfitt has argued, post-development scholars 'seek to dismiss the concept of development by defining it only by reference to its top down authoritarian form' (Parfitt 2002: 34).

Simplifying development in terms of 'social minorities and majorities' (Esteva and Prakash 1998) or in terms of 'the West and the Rest' (Escobar 1995) produces a logic that becomes highly dependent on aesthetic and relativist values about what constitutes legitimate and authentic social action (Parfitt 2002, and below). It also misleads the reader into thinking that the poverty being described in these texts was primarily – or even *entirely* – the result of the discursive strategies and structures established by the Western powers after 1945. Reviewing primarily the work of Escobar (1995) Nederveen Pieterse (2000) challenges post-development scholars for associating 'development' with 'Development,' the coordinated actions of a very large and diverse number of governments, international agencies, charities and NGOs, to achieve material improvements in human well-being (cf. Corbridge 1998; Parfitt 2002: 31). Within the same institution (in this case the World Bank), he argues, one can detect 'tremendous discontinuities' in development discourses over time and, one can add, among individuals and departments (Nederveen Pieterse 2000).

The account is 'totalizing' in a number of ways. First, it ascribes universal qualities to what is effectively a selective treatment of development policies and programmes in a single Latin American country (Colombia). As David

Lehmann (1997) has argued, Escobar's account underplays both the work of Foucault (see below) and other historical events and experiences (e.g. capitalist development in Northeast Asia) that would (potentially) contradict his selective rendering of 'development.' For Lehmann, Escobar's inability to provide a convincing or coherent account stems in large part from his 'parochial' reading of development, and from his inability to differentiate himself from the 'populism of the dependency generation … whose project he evidently wishes to replace' (Lehmann 1997: 577).

Along similar lines, Morgan Brigg (2002) argues that post-development (and here he is referring primarily to the work of Escobar 1995; and Esteva 1992) has 'tended to draw upon the more evocative aspects of Foucault's work on normalization,' emphasizing norms and discourses that exclude, as opposed to the ones that include. Reflecting upon Esteva's treatment (1992) of Truman's Point Four programme, Brigg suggests that by 'ascribing agency to Truman or to the Americans' (Brigg 2002: 425), post-development undervalues the role of negotiation, agency *and accident* in the determination of social outcomes, and underplays the possibility that the establishment of the development paradigm (which was never a static phenomenon) may have been the result of actors and forces outside of this homogeneous West. From this perspective, he argues, post-development scholars (and here we can certainly include Escobar [1995]) unwittingly undermine their own faith in alternative social formations by exaggerating – and *empowering* – the role of the United States in the establishment of the dominant development discourse. Far more useful, Brigg argues, is Foucault's understanding of bio-power, which suggests that power is exercised not by excluding 'social majorities,' but 'by assiduously integrating them into the regime of power' (Brigg 2002: 428). Central to this process is the classification of human subjects and nation-states in relation to a particular development norm:

> … (the) proliferation of writing and statistics renders the nation-states and human subjects of the Third World sufficiently visible that they may be distributed and evaluated against the norm. A wide range of disciplines and foci are relevant here, but the per capita GNP measure, with its capacity at once to totalize the field through the compilation of tables, thereby rendering nation-states and subjects visible while simultaneously finely differentiating them, is perhaps most exemplary of developmentalist power-knowledge.
>
> (Brigg 2002: 430)

Second, by suggesting that reality may be defined only or principally through discourse, Escobar is unable to define what constitutes 'real' suffering, alienation etc. and is therefore forced to embrace a relativism that supports all that global capitalism is not. As Parfitt (2002) has argued, Escobar shares with Foucault the notion that pre-modern society had an 'original' (and apparently, more desirable) condition that was historically undermined and usurped by the normalizing practices and discourses of modernity, and (in Escobar's case) of development. However, lacking a foundational statement about what constitutes the good life,

Escobar is ultimately forced to concede that material needs and realities do have an existence that is independent of the discourses we use to describe them (and people do experience material deprivation which may conform with discursive renderings of 'poverty'), and that development interventions may have addressed some of these needs but, in the grand scheme, the discourse and strategy of development only lead to 'underdevelopment and impoverishment, untold exploitation and oppression,' (Escobar 1995: 4).

For many, such 'wordplay' appears to trivialize the real and substantial achievements made by (and in the name of) development since the end of the Second World War. Citing improvements in life expectancy in India, for instance, Corbridge (1998) argues that not only is the failure to acknowledge these and other development achievements a selective interpretation of history, it is also unethical and dishonest. In his own words,

> This oversight is disgraceful and it cannot be justified by reference to the darker side of development. Development is about dilemmas, and the shortcomings of development should not be read as the Failure of Development. The so-called discourse of development is more critical and reflexive than its critics allow, and it is not fully determined by the powerful interest groups which figure in its production and dissemination.
>
> (Corbridge 1998: 145)

Misattribution also creates confusion about what leads to conditions of poverty and deprivation in the first place. Widespread in the post-development literature is the notion that development discourses (which include practices) created categories (such as illiterate, peasants, landless, squatters, etc.) that reified and/or created new forms of social exclusion. Although one can conceivably establish connections between certain discourses of development (e.g. neo-liberalism) and processes of material deprivation (e.g. inflationary spirals in the wake of devaluation), such associations are often left unspoken or simply defended on the basis of 'dogma and assertion' (Corbridge 1998).

Romanticism, relativism and representation

Implicit in the writing of Escobar (1995), Rahnema (1997a; 1997b) and Esteva and Prakash (1998) is the notion that people exploited by, on the margins of, or outside of the global capitalist system (Esteva and Prakash 1998), or the development paradigm (Escobar, 1995) or modernity (Rahnema 1997a; 1997b), are somehow living or in pursuit of a life that is more authentic, sustainable and ethical than that which they are explicitly, or by implication, in the process of resisting. As a number of critics (e.g. Peet and Hartwick 1999; Corbridge 1998; Nederveen Pieterse 2000; Parfitt 2002) have pointed out, such claims make distinctions that appear arbitrary and subject to the values and biases of individual authors (see below). They also belie an idealization of local and primarily rural forms of self-sufficiency. As Stuart Corbridge has argued, 'The key to the good life would seem to reside in simplicity, frugality, meeting basic needs from local soils, and shitting together in the commons' (Corbridge 1998: 142).

Beyond the arguments that can be marshaled against the utopian ways in which development is being construed in these instances (e.g. Corbridge 1998), the post-development of Rahnema (1997a; 1997b), Esteva and Prakash (1998), as well as Shiva (1997) and Nandy (1997) can be challenged for elevating on the basis of seemingly arbitrary principles the status of non-Western, non-scientific and anti-intellectual forms of action and experience (Corbridge 1998; Schuurman 2000; Nederveen Pieterse 2000; Lehmann 1997). For instance, Nederveen Pieterse (2000) challenges the simplistic notion that processes of exploitation, deprivation and, for that matter, 'global capitalism' are somehow unique to the West or to development. In his review of *Encountering Development*, David Lehmann (1997) suggests that Escobar's aversion (1995) to the role of science and professional expertise in the formulation and implementation of environmental policy constitutes an arbitrary and unfounded rejection of other legitimate worldviews (i.e. one which may be just as authentic – and possibly as effective – as those he appears to value).

The argument being made in these instances is not that science and the manifestations of science should not be subject to critical scrutiny (cf. Corbridge 1998; Forsyth 2003), but rather that science and the possible benefits of science are being dismissed on the basis of an ambiguous and ill-defined aversion towards the Enlightenment project (Peet and Hartwick 1999). Corbridge captures the essence of this important idea:

> At its best, the scientific method, or Enlightenment's reason, forces us to think that we might be wrong ... and so encourages an attitude of humility, a willingness to tolerate provisional truths, that is the very opposite of the arrogant certainties that post-development associates with Western science.
>
> (Corbridge 1998: 144)

The weakness and inconsistency highlighted by Corbridge (1998) and by others (e.g. Nederveen Pieterse [2000]; Schuurman [2000]; Parfitt [2002]) reflect a deeper relativism that pervades the work of many post-development scholars, including Escobar (1995) and especially Esteva and Prakash (1998). Underlying this ambiguity is an under- or non-developed articulation of the norms and criteria on which post-development scholars differentiate the movements, processes, ideas and ideologies they appear to favour in their own discourses from the ones they do not (cf. Corbridge 1998; Parfitt 2002). For Parfitt (2002), the lack of explicit and foundational assumptions creates two kinds of relativism. The first is that we are left with no logical means of distinguishing between good and bad forms of social action:

> If we can only know reality through discourse, what criteria are available to enable us to make truth claims in favour of one discourse (such as Escobar's post-development position) as compared with another (such as the development discourse)?
>
> (Parfitt 2002: 30)

The second is that it renders senseless the internal validity of the discourses being advanced by post-development discourse analysis. As Parfitt points out, '(Post-development theorists want) to cast the truth claims of the development discourse into doubt, but still made truth claims for their own discourse' (Parfitt 2002: 45).[56]

The dilemma thus created is that any calls for action or critiques of prevailing social discourses and orders are severely limited by the fact that they fail to subject to the same kinds of incredulity their own discourses and assumptions (cf. Lehmann 1997). The result is what is often portrayed as an empty shell of 'dogma and assertion' (Corbridge 1998), and one that fails to apply to its own discourse the (implicit) norms and standards it applies to the development discourse (Parfitt 2002). More pragmatically, a problem that arises as a result of the relativism that underlies the post-development critique is that post-development scholars appear to offer very little (beyond some very ambiguous calls for collective action on the basis of local needs and identity) in the way of alternative models or social movements (Nederveen Pieterse 1998; Corbridge 1998).[57] In Jan Nederveen Pieterse's words:

> Post-development is caught in a rhetorical gridlock. Using discourse analysis as an ideological platform invites political impasse and quietism. In the end post-development offers no politics besides the self-organising capacity of the poor, which actually lets the development responsibility of states and international institutions off the hook. Post-development arrives at development agnosticism by a different route, but shares the abdication of development with neoliberalism.
>
> (Nederveen Pieterse 2000: 187)

In short, post-development may be challenged for advancing an agenda that fails to articulate with sufficient clarity or purpose the ways in which their ideology, their methodology *or their politics* would move beyond the perceived failures of what was being described as the development orthodoxy (Nederveen Pieterse 1998, 2000; Lehmann 1997; Parfitt 2002; Corbridge 1998). Concerns were raised that anti-foundational theorizing would undermine a state whose ability to effect and explain revolutionary change was already the object of sustained neo-liberal scorn (Lehmann 1997; Nederveen Pieterse 1998, 2000). Concerns were also raised that post-development would foster the kinds of nihilistic *anti-development* (Nederveen Pieterse 2000) that questioned not only the dominant discourses and practices of the IMF, the World Bank, etc., but also called into question any effort to objectify or improve the human condition.

The following section considers a different form of discourse analysis, exemplified in the work of James Ferguson (1990 [1994]), whose study of bureaucratic power and development in Lesotho illustrates more carefully (and convincingly) the ways in which development discourses may determine what is visible and what is permissible in the context of development. In what follows, I shall argue that Ferguson's analysis sheds the kinds of essentialized rhetoric being advanced by

Escobar (1995), Esteva (1992) and (especially) by Esteva and Prakash (1998), and advances a more critical understanding of the ways in which the discourses of development projects and of 'development' may be used to legitimate and expand political and bureaucratic power.

Encountering James Ferguson

Ferguson's analysis (1990 [1994]) starts from the premise that the 'Lesotho' described in the 'official' development discourse of the World Bank, the Canadian International Development Agency and other international development agencies bears little resemblance to the Lesotho he finds in the economic history of the region (a literature which, he implies, is accurate and readily available to the professionals who write World Bank country reports). Far from being the 'traditional subsistence peasant society' the World Bank country report suggests it is, the 'academic discourse' reveals that Lesotho 'entered the twentieth century' as 'a producer of cash crops for the South African market' and an exporter of migrant labour (Ferguson 1990 [1994: 27]).

That the World Bank would so badly misconstrue the reality of Lesotho's economic history is not accidental or 'bad scholarship,' Ferguson argues. On the contrary, the rendering of Lesotho is intentional, and serves to legitimate the application of the kinds of technical and administrative solutions the World Bank (and the Government of Lesotho) has on offer. By erasing the history (and present reality) of colonialism and of 'politics,' Ferguson argues, the World Bank manufactures a problem to which it can legitimately respond. To legitimate this action, he suggests, it must also generate a discourse that defines what constitutes acceptable and necessary intervention in relation to the object of development (i.e. the 'Lesotho' of the development discourse).

For Ferguson, the central questions that stem from this characterization of Lesotho concern: first, the distinction between academic discourse and the official development discourse of international development agencies; second, the material and symbolic effects of the official development discourse; and third, the mechanisms by which this discourse 'maintains its own distinctive qualities,' (Ferguson 1990 [1994: 28]). The principal methodology is one that locates 'the intelligibility of a series of events and transformations not in the intentions guiding the actions of one or more animating subjects, but in the systematic nature of the social reality that results from those actions,' (Ferguson 1994: 18). For Ferguson, the question is, therefore,

> ... not 'how closely do these ideas approximate the truth,' but 'what effects do these ideas (which may or may not happen to be true) bring about? How are they connected with and implicated in larger social processes?'
>
> (Ferguson 1990 [1994: xv])

Like Escobar (1995), Ferguson (1990 [1994]) rejects the ontology, essentialism and interest-based functionalism of what he calls 'political economy,' and embraces instead an orientation that allows for identity, subjectivity, contingency and accident. His motivations for doing so are twofold. First, he argues, political economy imputes 'an economic function to 'development' projects,' and gives an impression that development is primarily or even only an effort to bring about 'a particular sort of economic transformation' towards capitalism, petty commodity production, etc. (Ferguson, 1994: 14). Second, he argues that political economy overstates the idea that social actions and institutions are always or essentially designed to serve the interests of particular social actors and forces. To avoid the imputed functionalism and essentialism of what he terms political economy, Ferguson offers a 'way of connecting outcomes with power, one that avoids giving a central place to any actor or entity conceived as a 'powerful' subject' (Ferguson 1990 [1994: 19]).[58]

Drawing directly upon Foucault's genealogy of the prison, Ferguson suggests that development institutions 'generate their own form of discourse, and this discourse simultaneously constructs Lesotho as a particular kind of object of knowledge, and creates a structure of knowledge around the object' (Ferguson 1990 [1994: xiv]). Framed in this way,

> ... discourse is a practice, it is structured, and it has real effects which are much more profound than simply 'mystification.' The thoughts and actions of 'development' bureaucrats are powerfully shaped by the world of acceptable statements and utterances within which they live; and what they do and do not do is a product not only of the interests of various nations, classes, or international agencies, but also, at the same time, of a working out of this complex structure of knowledge.
>
> (Ferguson, 1990 [1994: 18])

The production of discourse in Ferguson's view entails both a definition of what constitutes development success and failure, but also a 'representation of economic and social life which denies 'politics' and, to the extent that it is successful, suspends its effects' (Ferguson 1990 [1994: xiv-xv]). Such *depoliticization* is necessary, he argues, because it serves and facilitates the application, consolidation and expansion of bureaucratic power and planning that allow development to happen. By such 'unspoken' logic, he concludes

> ... the development apparatus in Lesotho is not a machine for eliminating poverty that is incidentally involved with the state bureaucracy; it is a machine for reinforcing and expanding the exercise of bureaucratic state power, which incidentally takes 'poverty' as its point of entry,
>
> (Ferguson, 1990 [1994: 255])

Discourse, agency and power

Many things can be (and have been) said about Ferguson's treatment of development. First, his characterization of development clearly sheds the kinds of essentialized rhetoric and moral outrage being advanced by Escobar (1995), Esteva (1992) and (especially) by Esteva and Prakash (1998). Second, his treatment of discourse reveals a relationship that is clearly more complex than that being provided by Escobar (1995).

Underlying Ferguson's treatment of discourse and power is the idea that discourse constitutes one (important) part of a larger machine, whose logic can be discerned only in relation to the intended and unintended outcomes of the project. In his own words,

> ... the structured discourse of planning and its corresponding field of knowledge are important, but only as part of a larger 'machine,' an anonymous set of interrelations that only ends up having a kind of retrospective coherence ...
>
> (Ferguson 1990 [1994: 275–6])

Drawing upon Foucault (1975 [1977], 1978 [1990]), Ferguson suggests that the structured outcomes of the Thaba-Tseka project are not necessarily the result of intentional planning, but that the project's activities and interventions 'can only be understood' in relation to the discursive regime 'that orders the "conceptual apparatus" of official thinking and planning about "development" in Lesotho,' (Ferguson, 1990 [1994: 275]). As one cog in the 'machine,' he argues,

> ... the planning apparatus is not the 'source' of whatever structural changes that may come about, but only one among a number of links in the mechanism that produces them. Discourse and thought are articulated in such a 'machine' with other practices ... but there is no reason to regard them as 'master practices,' over-determining all others.
>
> (Ferguson 1990 [1994: 275–6])

Ferguson therefore ascribes to the discursive practices of development agencies a power that produces the rationality of what constitutes successful and unsuccessful forms of development. However, as Johaan Graaff (2006) has pointed out, Ferguson's analysis also ascribes deep meaning and causality to a system whose reality we can divine only through its effects. The logic, he argues, 'is not apparent to the people who constitute the machine, but nevertheless the machine has regular "side-effects," which indicate its "intelligence"' (Graaff 2006: 1392). But 'if "ordinary" acting individuals caught up in the coils of the machine are unable to fathom its "retrospective intelligence,"' he asks, 'how can social scientists?' (Graaff 2006: 1393–4). Moreover, 'if the impact of discourse is so universal and so powerful, how do intellectuals not also succumb to its pernicious influence?' (Graaff 2006: 1394). The implication, Graaff suggests, is that reality is defined and interpreted in relation to 'subject-less structures with an intellectual who then defines their logic' (Graaff 2006: 1393).

Construed in this way, Ferguson's treatment suggests a functionality in which the logic and importance of any particular outcome or effect can be explained only in relation to the contribution it makes to the larger whole or 'machine.'

Although he rejects the idea that discourses and practices are the equivalent of 'master practices,' which over-determine all other forms of behaviour, it is difficult to avoid in Ferguson the conclusion that the agency of individuals is essentially the product of wider forces subject to the logic of 'the machine' and (in Graaff's estimation) to the ontological assumptions and biases of social science researchers. As Johaan Graaff has argued, the scenario being described is one in which,

> Third World countries (are) reduced to helpless victim status by an unstoppable development discourse embodied in the all-powerful dispositif of the World Bank, IMF and WTO.
>
> (Graaff 2006: 1394)

Similar assertions can be made about Ferguson's treatment of 'development.' For instance, Stuart Corbridge *et al.* (2005) challenge Ferguson (1990 [1994]) for his failure to explore the aims, motivations and actions of the development projects he documents in his study. Although Ferguson goes to great lengths to represent the perceptions of the program's intended beneficiaries, they argue, he 'fails to extend the same courtesy to the aid workers he spent time with. Oddly, for an anthropologist, these men and women are not allowed to speak for themselves' (Corbridge *et al.* 2005: 259). Along similar lines, David Mosse argues that Ferguson fails to capture 'the complexity of policy making and its relationship to project practice, or to the creativity and skill involved in negotiating development' (Mosse 2005: 2).

To counter the notion that development is only or essentially inclined towards the expansion of bureaucratic power, more recent work has begun to explore the ways in which aid bureaucracies, government agencies and beneficiaries negotiate and construct the discourses that structure and give meaning to development policies and practice. For instance, Stuart Corbridge *et al.* (2005) use Foucault's concepts of biopolitics and governmentality to understand the ways in which governments and government beneficiaries perceive poverty and employment programmes in rural India. 'Biopolitics,' they argue, 'refers to those government interventions that seek to improve the quality of a population as a whole, and these procedures produce that which we name the state as the effect of these interventions' (Corbridge *et al.* 2005: 15). 'Governmentality,' on the other hand, involves

> … a further extension of powers to those who profess expertise over the private body or the body public, be they aid workers, economists, psychologists, psychiatrists, social workers, sexologists or public health workers.
>
> (Corbridge *et al.* 2005: 16)

Like Ferguson (and Escobar), the authors start from the proposition that discourses of assistance, responsibility and expertise can define what is knowable, what is permissible and what is necessary in the context of development. However, their rendering of governmentality and of development is not one that ascribes unseen power and interest to the unstated actions of government policies and projects, but one which starts from the assumption that 'the state' is constructed and legitimated through many different actions and discourses, and that these discourses may be unravelled and exposed through careful conversation with the agents and beneficiaries of development. Towards this end, they devote considerable time and resources to the difficult task of documenting the different ways in which beneficiaries and lower-level government officials understand and interpret the power of the Indian state(s).

Along similar lines, David Mosse (2005) uses a British aid programme in western India to understand the ways in which policy discourses may be used to mobilize and legitimate political support in the face of project 'failures' and changing donor needs and priorities. Drawing upon more than 10 years of personal involvement in DFID's Indo-British Rainfed Farming Programme, Mosse (2005) documents the ways in which the project's central aims and priorities changed from what was first construed as a rainfed farming project to a watershed conservation project to what would finally become a rural livelihoods project. According to Mosse, policy discourses derive their power not from the rules and incentives they provide for future action, but from their 'vagueness, ambiguity and lack of conceptual precision,' which are '*required* to conceal ideological differences so as to allow compromise and the enrolment of different interests and to multiply the criteria of success within project systems' (Mosse 2005: 230). Mosse's principal claim is that development projects thrive and survive on the basis of their ability to conform to a coherent set of policy ideas, and to maintain the organizational networks and relations that support and identify with these ideas. Framed in this way, projects can never fail: 'they are failed by wider networks of support and validation,' (Mosse 2005: 18).

Given his long period of involvement with the project, Mosse is able to assemble what is, comparatively speaking, an unusually large amount of 'inside' information and insight (including internal project documents, reports, consultations, conversations and evaluations) about the various actions, conflicts and events that led to the discourses he describes in his study. The combination of access and praxis, therefore, offers important insights about the ways in which discourse may be used to understand the acts and assumptions of development. However, as we shall see, it also splits a very fine line between a professional discourse which defines the norms upon which individuals may act within the 'development community' and an academic discourse that aims to question assumptions and offer alternatives, no matter how 'practical' these may be.

In the following section we consider the challenge of combining critical and professional discourses of development and social change.

Policy, discourse and praxis

In his *Epilogue* to *The Anti-Politics Machine*, Ferguson offers a *post-hoc* response to the notion that his analysis fails to provide a prescription or guide that would address the de-politicizing processes he describes in his study. Beyond the standard line that 'making the world a better place' was not the central aim of the book, Ferguson shares these concerns, and asks of himself, and of others engaged in development research, the following question:

> How can we work for the social and economic changes that would make a difference for the ordinary people we have known as informants, neighbors and friends?
>
> (Ferguson, 1990 [1994: 285])

On this question, and in contrast to Escobar (1995), Ferguson aims to distance himself from the idea that scholarship can or should be connected to an explicit political agenda, and suggests instead that efforts to change the world (potentially on the basis of social science scholarship) may be affected (and adversely distorted it would seem) by the agents (multinational agencies, national governments, etc.) that try to act or speak on behalf of the poor. Starting from the premise that the interests of governments and government elites 'are not congruent with those of the governed,' Ferguson (1990 [1994: 280]) suggests that any effort to advance the interests of the poor is 'worse than meaningless' in the sense that 'it acts to disguise what are in fact highly partial and interested interventions as universal, disinterested, and inherently benevolent.'

Although Ferguson's conclusions are fully consistent with the arguments he advances in the main body of the text, one is left wondering whether the 'concern' being articulated in this instance is that development efforts fail to represent the needs of the poor or whether they are not entirely authentic, or altruistic. If the latter, one is tempted to conclude that Ferguson is perhaps over-stating the ability of (any) development project to operate purely on the basis of altruism (cf. Brett 1993). He also gives the impression that the only good (or authentic) development interventions are ones that advance *only* the interests of those in need (as opposed to those doing the developing). By demonizing efforts to 'disguise' the 'real' intentions of development, Ferguson therefore denies the possibility that politics may influence the ways in which the masses (or particular groups within the masses) may influence government (including government elites) and the ways in which government orients itself towards the masses.[59] In so doing, he squares the circle of a logic that is deeply skeptical about the ability of government or of society to advance the 'real' interests of those in need. Such a conclusion raises difficult questions about the utility of scholarship, questions Ferguson only partially addresses by suggesting that scholars can and should be politically engaged in their own societies, and in their respective areas of study.

But the solutions Ferguson presents us with are also deeply problematic. Assuming that governments and government 'elites' will never act in the interests of their citizens (or more ambiguously, they will only act in ways that also happen to serve their own interests), the best we can do is to let people get on with their own lives, and to avoid the tricky business of politics and development altogether. Ferguson suggests that the ability of development interventions to represent the poor is complicated by the fact that 'the people' and the poor 'are not an undifferentiated mass,' (Ferguson, 1990 [1994: 281]), but are in reality a diversity of many different lives replete with different needs, interests and identities. Ferguson then comes as close to Von Hayek as his ethnographic sensibilities will allow:

> There is not one question – 'what is to be done?' – but hundreds: what should the mineworkers do, what should the abandoned old women do, what should the unemployed do, and on and on. It seems, at the least, presumptuous to offer prescriptions here. The toiling miners and the abandoned old women know the tactics proper to their situations far better than any expert does. Indeed, the only general answer to the question, 'What should they do?' is 'They are doing it!'
> (Ferguson 1990 [1994: 281])

Although it would be misleading to suggest that Ferguson (or Foucault for that matter) was essentially a neo-classical wolf in postmodern clothing, it is difficult to avoid the conclusion that the implications he raises in his analysis are, at the very least, highly consistent with the neo-classical trope. Compare the preceding quotation, for instance, with this one, by William Easterly:

> Planners announce good intentions but don't motivate anyone to carry them out; searchers find things that work and get some reward. Planners raise expectations but take no responsibility for meeting them; searchers accept responsibility for their actions. Planners determine what to supply; searchers adapt to local conditions. Planners at the top lack knowledge of the bottom; searchers find out what the reality is at the bottom. Planners never hear whether the planned got what it needed; searchers find out if the customer is satisfied.
> (Easterly 2006: 5–6)[60]

Ferguson's *Epilogue* draws our attention to the ways in which researchers legitimate their own work in relation to the norms and practices of academic sub-fields and disciplines, and to the moral terms on which academic researchers extract time, information and other forms of data from their informants in the field (whose opportunity costs in this case are extremely high). Through teaching, public speaking and advocacy, Ferguson suggests, academics may '(apply) their specialized knowledge to the task of combating imperialist policies and advancing the causes of Third World peoples.' (Ferguson 1990 [1994: 286]) Strategies of this kind may be made relevant, he argues, by criticizing or informing governments in donor and in Third World countries, and by forming alliances with other points of 'counter-hegemonic' engagement, such as labour unions, opposition political parties, peasants' organizations, churches, etc. Although he is careful to add that 'counter-hegemonic' can be defined only in relation to what is deemed relevant and hegemonic in a particular 'local' context, it is clear that the praxis he is advancing in this instance is one in which academics support struggles and movements that are defined primarily in relation to all that global imperialism (and other forms of hegemony) is not.

However, the notion that 'academics' (or anthropologists for that matter) would automatically align themselves with a struggle against global capitalism/imperialism or in favour of Third World 'causes' is in a number of ways problematic. First it implies that 'academics' would – or should – be inclined to engage themselves in struggles of this kind and, second, that the struggles thus produced would easily or unproblematically reflect 'the causes of Third World peoples.' As Corbridge *et al.* (2005: f/n 2, p. 266) point out in their own *Postscript*, well-meaning anthropologists are not the only academics whose work aims to understand and affect policy in the developing world. Equally or even more influential, they suggest, are the 'activists of the New Right,' whose emphasis on neo-classical economics and neo-liberal policy reform are indeed part of the hegemony Ferguson and other post-development scholars have in mind. On this question, Ferguson provides little detail about the ways in which academic and/or other social movements would represent the real needs and 'causes' of Third World peoples, and/or how 'Third World peoples' would make their voices heard in this process.

Ferguson's *Epilogue* draws our attention, finally, to the difficult (and somewhat arbitrary) distinction he makes between 'academic' and 'development' discourse. Underlying his argument in favour of political engagement is the notion that academics make their findings relevant to the individuals and institutions that decide social and economic policy in the developing world. By providing first-hand accounts of Salvadorian death squads, real-life Palestinians or 'the realities of South Africa,' Ferguson (1990 [1994: 286]) suggests that the anthropologist (and by implication, the academic) 'has both an opportunity and a responsibility to enter into the political debates' about *apartheid* and other matters relevant to their research. However, the argument being made in this instance gives the impression that the individuals and institutions that determine policy are equally open and receptive to any 'truth' the anthropologist may be able to unravel. 'Anthropologists come cheap,' Ferguson (1990 [1994: 287]) suggests, but can the same be said about their ideas and ideals? If academics engage in discourses that truly challenge intellectual and geo-political hegemonies, is the relationship between academic inquiry and policy discourse really so straightforward?

To answer this question, we need only look at the Preface of *Cultivating Development*, by David Mosse (2005), in which the author describes in great detail the extent to which the UK Department for International Development took 'strong exception' to the findings of his decade-long involvement in its Indo-British Rainfed Farming Project. Responding to what it felt was an unfair and inaccurate rendering of its development work, DFID tried to persuade the author's university, publisher, professional association and head of department to suspend publication of the manuscript. Although it was ultimately unsuccessful, the response by DFID helps to illustrate with great clarity the problems that can arise when academics challenge the intellectual and institutional 'hegemony' of a particular agency or project.[61]

Concluding remarks

This chapter has explored the ways in which Foucauldian discourse analysis may be used to understand and advance processes of economic development and social change. Reviewing first the work of Esteva (1992), Escobar (1995), Rahnema (1997a; 1997b) and Esteva and Prakash (1998), I argue that post-development writing about development tends to engage in a 'meta-narrative' of exploitation and salvation, in which efforts to improve the life prospects and living conditions of the world's poor is associated with the normalizing power of large aid agencies and the West. As we have seen, the discourse is in many ways totalizing, and problematic. For one, the notion that poverty is constituted through discourse gives the impression that poverty, hunger, health, pain, etc., are only or primarily social constructions, trivializing the true plight of the poor and simplifying the important ways in which development efforts have improved the quality of life (e.g. by eliminating smallpox) through development discourse. Second, the notion that discourse is imbued with power and interest slips too easily into a relativism that castigates all that is 'evil' (Rahnema 1997b: 400), which for many post-development authors appears to be anything that resembles what is essentially a crude caricature of capitalism, modernization and 'the West.'

Therefore, what can we conclude about the viability and desirability of using post-development to understand and advance a more critical approach to development? First, the scholarship quite clearly aims to recognize and give voice to those undervalued, unrecognized and *compartmentalized* within dominant development discourse. Second, it does so in a way that situates development in a wider historical context of globalization, modernization and geo-political change. However, as the preceding suggests, it fails to articulate with sufficient clarity or purpose the ways in which its vision of local self-sufficiency and global solidarity would challenge the dominant development paradigm. Moreover, its rejection of foundationalism leads to a relativism that glorifies any movement or idea that rejects what is in effect a gross simplification of globalization, capitalism and development.

I suggest that James Ferguson's study of bureaucratic power and development in Lesotho is far more coherent. Unlike Escobar (1995) and Rahnema (1997a; 1997b), Ferguson avoids (until the *Epilogue*) the kinds of moral outrage and essentialized rhetoric being advanced by post-development, and offers instead a more modest assessment of the ways in which development discourses may be used to legitimate the application and expansion of bureaucratic power. In so doing, he draws our attention to the ways in which words and language may be used to classify and define what is permissible, what is knowable and what is necessary in the context of development.

However, Ferguson also has difficulty reconciling the 'development discourses' he associates with the various documents and country reports of CIDA and the World Bank, and the 'academic discourse' of which he is a part. For one, his representation of the academic discourse is at times naively uncritical about the historical truth claims of the 'academic' literature on Lesotho. Although the lines between the World Bank country reports and the scholarly literature Ferguson describes are different enough, one is tempted to wonder whether the distinctions he is making are either over-stated or reflective of a time (in this case the 1960s and 1970s) when development research and academic scholarship were more clearly differentiated than they are today.[62] Whatever the case, the distinction Ferguson suggests is a body of scholarship whose rendering of economic history and of Lesotho is accurate and largely unproblematic and a development discourse whose interpretation is fundamentally biased in favour of instrumental needs and effects. Although he does a masterful job of deconstructing the discourse of the World Bank, CIDA, etc., one is left with the impression that the academic discourse he uses to 'expose' the development discourse is in some way immune from the kinds of bias and influence he associates with the development discourse.

Second, and more relevant to the scholarship that would follow, Ferguson has great difficulty reconciling the observations he makes about power, exploitation, etc., and his own role as a scholar. In the *Epilogue*, he makes the case that scholars can make their research and scholarship relevant by participating in policy dialogues about human rights abuses, genocide, and so on. However, the model he suggests understates dramatically the kinds of power that can influence policy processes and agendas. Indeed, one could make the case that his model of research, scholarship and policy is not at all inconsistent with the kinds of 'instrumental' models of rational decision making described by Mosse (2005). Far more engaged with the subject of his research is David Mosse's ethnography of aid.

Mosse's study raises a number of difficult questions about the desirability and viability of using discourse analysis to understand and advance processes of development and social change. First, it quite clearly moves beyond the notion that development discourses are only or essentially inclined towards the expansion of power, and suggests instead that policy discourses have the effect of creating or reducing power, but that they are also affected by real actors, decisions and events. Second, it provides a body of evidence that allows the reader to interpret the ways in which different actors and interests actually shape (and re-shape) policy discourses. Third, and in contrast to Ferguson, it connects the findings of this research with an explicit political agenda.

Mosse's experience (2005) also helps to illustrate the kinds of political and epistemological struggles that can arise when scholars find themselves on the 'wrong' side of what a development bureaucracy deems 'acceptable' policy discourse. Unlike positivism and unlike many forms of (scientific) Marxism, the locus of power in discourse analysis rests not in the artificial confines of scientific inquiry and/or modes of production but in the discursive realm of written words and contested histories, a realm in which scholars are often (wittingly and unwittingly) forced into a more active and critical self-awareness about the impact of their own research and analysis on development policy and practice.

The following chapter explores in more detail the ethical and procedural challenges of representing 'the other' in social science scholarship (Kapoor 2004; Corbridge 1994) and of converting these insights into what is deemed legitimate or useful forms of knowledge. As we shall see, the idea of advancing an interpretive approach to the study of development would become increasingly attractive to development researchers during the 1990s, whose emphasis on participatory research methodologies would shift the locus of change from class relations and class struggle to the 'ontological encounter' between the subject and object of development research.

5   Development as freedom of choice

From measurement to empowerment to rational choice

> It's all about choice, graduates, it really is. There is nothing about fate. It's all about values, creativity, leadership … One day's Pentagon spending would provide all sleeping sites in Africa with five years of bed net coverage to fend off a disease which kills millions every year. That's a choice. We haven't made it.
>
> Jeffrey Sachs, Commencement Speech to the Graduating Class of 2007
> Ursinus College (Quoted in *The New York Times*, 10 June 2007)

Introduction

The last two chapters have explored the dominant efforts within Marxism and postmodernism to theorize and advance an alternative to neo-classical theory. In neither case have we found very much reason for optimism. This chapter explores the seemingly counter-intuitive idea that liberalism may provide an alternative to the neo-liberal/neo-classical frame.

At first glance, the comparison seems dubious. Insofar as it supports and expands the ability of individuals to make basic decisions concerning the investment of time and money and the consumption of goods and services, neo-liberalism embodies the liberal idea that the good society is one in which individuals enjoy maximum freedom to decide the choices and values that have bearing in their lives. However, the assumption that social welfare and distribution may be determined only or primarily on the basis of individual decision making and market forces dramatically underplays the traditional focus within liberalism on the moral and distributional consequences of individual decisions and choices (Sen 1999 [2001]).

In its broadest sense, liberalism aims to reconcile the use of power within society with the autonomy and well-being of the individual. Beyond the aim of protecting individuals from arbitrary and abusive power, a central issue that has long preoccupied liberal theory is the challenge of establishing the normative conditions under which individual liberty may be reconciled with competing ideas about what constitutes the public good. John Stuart Mill's 'harm principle,' for instance, suggests that individuals may be free to do as they please so long as their actions do not infringe upon the utilities and liberties of others. In *A Theory of Justice*, John Rawls (1971) imagines an 'original position' in which the freedom to choose entails the establishment of social arrangements grounded in the re-distribution of individual liberties. Finally, Robert Nozick (1974), whose *State, Anarchy and Utopia* represents what is probably the most radical (libertarian) argument in defense of individual freedom and private property allows for certain 'catastrophic moral horrors,' on the basis of which basic liberties could be suspended.[63]

Liberalism also advances the idea that the good society is one in which individuals (and, by extension, societies of individuals) have the freedom to engage in self-determination. Reflecting on the criticisms advanced by postmodernism, Charles Taylor (1991), for instance, has argued that one's understanding of self-identity occurs in dialogue with others, and against a 'horizon of significance,' which structures social behaviour, but is also structured by inter-subjective communication. Postmodernism, he argues, (and here he is referring specifically to Derrida and Foucault) stresses the 'creative constructive' elements of social discourse, but forgets about the horizons and dialogical encounters that give these elements their meaning (Taylor 1991). Similarly, and perhaps most ambitiously, Jurgen Habermas theorizes an 'ideal speech situation' in which individuals are at their core able to 'overcome their at first subjectively biased views in favor of a rationally motivated agreement' (Habermas 1987: 315). Framed in this way, the ability to communicate inter-subjective norms, values and experiences has intrinsic value, defining essentially what it means to be human (Habermas 1987; Flyvbjerg 2001).

From this 'wider' perspective, the worldviews and policies put in place by neo-liberalism and neo-classical theory appear to abandon the traditional liberal concern for social welfare and the public good (cf. Corbridge 1993). Amartya Sen (1999[2001]), for instance, has argued that neo-classical theory fails to theorize sufficiently the normative conditions under which resources may be distributed within society. By making explicit the need for public discourse about what constitutes personal and social utility, Sen divorces himself quite consciously from the neo-classical notion that values may be established and goods and services distributed on the basis of supply and demand. Instead, he advances the idea that politics and public discourse must play a 'constructive' role in both the evaluation of basic social norms (e.g. governing free market economies), and the principles upon which commodities may be produced and distributed within an economy.

Sen's understanding of democracy and of freedom demands an open society that allows the opportunity for public discussion, and dissent. But how do freedoms of expression and association arise in the first place? And what form do they take?

This chapter now explores three 'innovations' that have aimed to conceptualize and capture an image of development that is less beholden to the formal assumptions of neo-classical theory. One is Sen's theory of capabilities and entitlement, which aimed to forge a theoretical path between the more deterministic varieties of historical materialism and the methodological individualism of neo-classical economics. A second and related area is the participatory research methodology of Robert Chambers. A final area of focus is what we may loosely call 'livelihoods theory,' an analytical framework that was developed by the UK DFID and the United Nations Development Program (UNDP) to put people 'at the centre of development, thereby increasing the effectiveness of development assistance' (DFID 1999: 1.1).

In what follows, I shall argue first that Sen's conceptualization of capabilities and entitlement helps to expand neo-classical assumptions about utility, rationality and value, shifting the intellectual emphasis away from an under-socialized model of individual decision making and rational choice to one rooted more substantially in a context of history, power and freedom. However, Sen's analysis also leaves open to interpretation the empirical and normative conditions under which basic rights of freedom and entitlement would be respected and accommodated within society. As a result, his defence of individual liberties and freedom at times slips into a logic of individual decision making and rational choice, highlighting the ideological and methodological importance of adopting or at least speaking the language of neo-classical discourse (cf. Gasper 2000).

By linking Sen with the participatory methodologies of participatory rural appraisal (PRA) and the sustainable livelihoods approach (SLA), I suggest that a conceptual and a causal link may be made between Sen's theory of capabilities and entitlement and what would become PRA and the SLA. Underlying PRA and the SLA is an assumption that participatory research methodologies would yield more accurate insights about local realities, which would in turn inform better policy. To achieve these ends, the SLA combines (with some difficulty) the methodological individualism of neo-classical theory and the notion that individual and household decisions may be constrained by a variety of social, cultural and political factors. Central to the framework is:

- an *ontology* that embraces the concepts of micro-economic theory (i.e. individuals, households, scarcity and exchange) to document the ways in which individuals and households adapt their livelihood portfolios in the context of environmental and macro-economic change;
- a *methodology* which emphasizes the collection of local contextual knowledge and first-hand accounts;
- an *epistemology* that makes a strong link between participatory methodology and better policy.

First, however, we explore in more detail Sen's conceptualization of freedom, and the conditions under which basic rights of freedom may be respected and preserved.

Poverty as 'capability deprivation': theorizing the work of Amartya Sen

Conventional approaches to the study of poverty aim to convert bits of information (what Kanbur and Shaffer [2006] have called 'brute data') into numerical representations, obtained primarily through household surveys. How one defines 'data' and what one does with it, of course, varies widely. However, within neo-classical economics the 'gold standard' (Kanbur and Shaffer 2006) for most researchers, governments and international agencies is the income and expenditure survey, which documents and calculates (through various means) the total amount spent by a household (either in money or in kind) on consumption of basic and other needs. Although the objects of inquiry may vary (with season, sector, demographics, research questions, etc.), the observations derived are assumed to provide an authoritative statement about the nature and extent of economic activity (i.e. income and expenditure) in a given locality. Underlying this assumption is the notion that pre-determined survey questionnaires and standardized sampling techniques will yield a methodology that is reliable (and therefore comparable) in different empirical settings. [64]

'Biological' approaches to the study of poverty aim to measure poverty by calculating the total income required 'to obtain the minimum necessities for the maintenance of merely physical efficiency' (Sen, 1981: 11, citing S. Rowntree). The World Bank's (World Bank 2000) self-declared approach to poverty assessment, for instance, uses income and expenditure data (obtained primarily from sample survey and price data provided by national governments) to calculate aggregate levels of consumption. To measure and compare poverty, the Bank uses national and purchasing power parity estimates, respectively, to calculate country-specific and global poverty lines, below which consumption is associated with poverty.

The value of measuring poverty in this way is that it provides an objective numerical standard upon which the performance of an economy may be assessed and compared. However, as Sen (1981: Chapter 2) has argued, biological approaches suffer from a number of problems, not least the difficulty of establishing (across cultures, lifestyles, metabolism and so on) what constitutes a 'minimum necessity,' associating nutritional needs and intake with what one actually consumes and finally, the difficulty of measuring *non-food items*, such as shelter, access to education, freedom etc. We shall return to these points in due course.

Sen's re-conceptualization of 'conventional' poverty assessments stems from his critique of utilitarian and liberal assessments of social welfare and individual liberty. For Sen, utilitarian theories place excessive normative weight on the kinds of social choices and arrangements that give the greatest amount of utility/welfare to the greatest number of people within society. By ranking the sum total of utility within society, neo-classical economics aims to evaluate the *consequences* of individual and collective choices and the *welfare* generated as a result of these choices. However, as Sen points out, utilitarian approaches say and do very little about the ways in which resources are actually distributed within the aggregate. Clearly, he concludes, any effort to evaluate social welfare must come to grips with questions of power, inequality and distribution.

To adequately address questions of distribution, Sen argues that welfare and value must be assessed not only by ranking the sum total of individual preferences, but also by taking into consideration the capabilities that each and every person has to establish and pursue these preferences. Take, for instance, the distinction he makes between his understanding of capabilities and the idea that poverty and well-being may be understood as a kind of surplus or deficit of 'human capital.' In contrast to the 'human capital approach,' which 'tends to concentrate on the

Perhaps the most important application of Sen's capabilities approach is the Human Development Index (HDI), which is published annually by the UNDP. Pioneered by Sen, the American legal ethicist Martha Nussbaum, and the late Mahbub ul Haq of the UNDP, the HDI aims to assess the performance of countries around the world by ranking them in relation to a set of human development indicators. These indicators include 'conventional' measurements, such as life expectancy, child and maternal mortality, literacy, income and GDP per capita, as well as ones that affect the capabilities that people living in these countries have in relation to major social groupings, such as gender, ethnicity and class. Included in this latter set of indicators are efforts to quantitatively rank access to things like clean drinking water, primary education, universal healthcare and public and private spending on education, healthcare and infrastructure.

'A person's capability,' Sen (1999 [2001: 75]) argues, 'refers to the alternative combinations of functionings that are feasible for her to achieve.' 'Functionings,' in turn, are 'the various things a person may value doing or being.' Whether or how a person is able to realize these functionings depends in large part upon a variety of personal, social and environmental factors, including health, age, illness and disability (what Sen calls 'personal heterogeneities'), variations in rainfall, temperature, the presence of infectious diseases ('environmental diversities'), 'variations in social climate' (governing, for instance the provision of public education, rates of crime, etc.), 'differences in relational perspectives' (covering social norms of reciprocity and welfare, and the social conditions under which individuals may 'appear in public without shame') and, finally, distribution within the family (Sen, 1999 [2001: 70–1]).[65]

Sources: UNDP (2000, 2004); Sen (1999 [2001])

Box 5.1 Capital and capabilities: the Human Development Index

agency of human beings in augmenting production possibilities,' Sen argues, the capabilities approach emphasizes 'the ability – the substantive freedom – of people to lead the lives they have reason to value and to enhance the real choices they have' (Sen, 1999 [2001: 293]). Using the example of formal education, he argues that the capabilities approach emphasizes not only the productive assets a formal education provides, but also less tangible benefits, such as the enjoyment of 'reading, communicating, arguing ... being taken more seriously by others, and so on' (Sen, 1999 [2001: 294]). The distinction is an important one, and one that draws our attention to the notion that capabilities, entitlement and poverty are necessarily rooted in the material and symbolic power of human relationships.

Bridging structure and agency: Sen's theory of entitlement

Like Marx, Sen suggests that a critical source of productivity and well-being in an economy stems from the organization of human labour. Whether and how individuals, families and other collectivities may turn their endowments into entitlements depends upon the norms that govern exchange conditions and production possibilities, as well as the various kinds of entitlement strategies – or relations – people may choose to pursue. In *Poverty and Famines*, for instance, Sen (1981: 2) identifies four entitlements:

- *trade-based entitlement*: 'one is entitled to own what one gets by trading something one owns with a willing party,';
- *production-based entitlement*: entitlements derived from the productive use of 'one's owned resources,' including resources hired from a willing party;
- *own-labour entitlement*: in a free market economy, 'one is entitled to one's own labour power,' which one may use to obtain trade- or production-based entitlements;
- *inheritance and transfer entitlement*: 'one is entitled to own what is willingly given to one by another who legitimately owns it.'

Sen's conceptualization of entitlement therefore marries together theoretical insights about production, consumption and the capabilities and constraints facing individuals in different personal, social and environmental circumstances. It also allows space for the ways in which people may value (and may be able to value) the lives they want to lead. For Sen, entitlement represents the bundle of commodities over which a person may establish effective and legitimate command in a market economy. One's personal or shared entitlement may reflect one's endowment (productive resources and assets of wealth which command a price in the market), as well as the 'production possibilities' (i.e. available knowledge and technology) and the 'exchange conditions' under which people buy and sell goods and services, and establish prices, in a market economy (Sen, 1999 [2001: 162–3]).

However, Sen is by no means suggesting that social outcomes are *only* the result of individual decisions or rational choice. For instance, in *Poverty and Famines* (Sen 1981) he makes the case that destitution leading to famine was the result of a combination of environmental factors and structural forces, arising as a result of a systemic commitment (on the part of the state, colonial authorities, etc.) to private property and liberalized trade. In so doing, he draws our attention to the ways in which systemic fluctuations (e.g. drought, inflation) and an ideological commitment to the principles of free market capitalism can set in motion multiple spirals of poverty leading to destitution.[66]

Central to Sen's understanding of entitlement is therefore a concept of legitimacy, which is connected necessarily to the norms and assumptions of a market economy. Formal or informal, rights of ownership and private property are embedded in a culture that enables individuals to use a particular 'commodity bundle' (Sen 1981), obtain benefits from using it, and (significantly) exclude others from using or benefiting from it, or from reducing its value (Gore 1993; Devereux 1996; Sen 1981).[67]

Sen's theory of public action

An important aspect of Sen's treatment of entitlements is the space he gives to the actions taken (and not taken) during times of hardship and crisis. For instance, in his analysis of the 1943 Bengal famine (Sen 1981: Chapter 6), Sen is concerned not only with the factors that triggered the famine (i.e. a failed harvest resulting from an unusually devastating cyclone, the Japanese occupation of Burma and a resultant downward pressure on the purchasing power of primarily agricultural labour) but also with the ways in which authorities (at many different levels) responded to the crisis. In the (relatively chaotic) context of spiraling food prices, panic hoarding and 'vigorous speculation,' Sen (1981) argues, was a colonial administration that was both unwilling and unable to perceive and therefore respond to the situation at hand. Part of this inaction reflected a refusal on the part of the British government to re-allocate shipping resources and to allow food imports into India during the Second World War. But an equally important factor, and one which explains the government's 'misreading' of the crisis, was a 'theoretical' and ideological worldview that shaped its understanding and treatment of the problem. Particularly influential at the time was the idea that famines could be predicted and managed by estimating the total availability of food in relation to existing demand requirements. Such an approach, Sen (1981) argues, underplayed dramatically the ways in which shifting exchange entitlements (i.e. falling incomes in the context of food inflation) would affect the ability of people to command food in a 'free' market economy.[68]

Sen's insights about famine, poverty and deprivation clearly influenced his thinking about the 'instrumental' and 'protective' importance of rights, freedom and democracy. His treatment of China's 'Great Leap' famine (Sen 1999 [2001]), for instance, suggests that freedoms of expression and association would prevent the kinds of action *and inaction* that led to the Chinese famine (in which an estimated 30 million people died of starvation and disease), and others like it. Citing post-independent India as an example of successful famine prevention, he argues that the widespread recognition of freedoms of expression, association and democratic participation help to account for the fact that no major famine has occurred in India since 1943. In contrast, the experience of the Great Leap famine, he argues, illustrates the ways in which efforts to enhance economic productivity on the basis of hierarchy, secrecy and authoritarianism can very easily lead to disasters of monumental proportions. Particularly important in his assessment of the Chinese famine were the incentives of local and regional officials within the Communist Party apparatus; wanting to please their superiors – and not wanting to admit failure – officials routinely concealed crop failures in the agricultural regions of western China. Lacking a free press and freedoms of expression and association, affected populations had no means by which they could take their claims and grievances to the central government; officials in the central party apparatus were therefore unaware of the disaster until it was far too late (Sen, 1999 [2001]).

Sen's analysis draws our attention to the political and historical conditions under which the freedom to decide major moral questions about the sanctity of private property, about the distribution of essential goods and about the protection of vulnerable populations may arise. His theory of famines, capabilities and entitlement also emphasizes the complex ways in which systemic factors (such as a commitment to liberalized trade) may constrain the choices available to poor and vulnerable groups in society.

Restricting the freedom to choose: Sen's theory of 'social commitment'

For Sen, the good society is, throughout his writing, one that establishes and defends through public discourse the freedom to discuss and debate what constitutes the good life, and to agitate societies, governments and other forms of authority when things go wrong. A crucial dimension of Sen's capabilities approach concerns the means by which individuals and societies may ascertain and evaluate the relative importance of what each individual deems important and desirable in his or her life. Reflecting upon 'the impossibility' of making inter-personal comparisons of what each individual has in mind when ranking and ordering individual preferences, Sen suggests that individual preferences may be ordered, but that they must take into account 'the heterogeneity of factors' (including health, disability, climate, etc.) that influence one's functioning and capabilities. It is here, Sen argues, that the freedom to choose (what he calls 'social choice') becomes vital. Dismissing the (postmodern and methodological positivist) notion that inter-personal comparisons are 'impossible,' Sen (1999 [2001]) argues that 'explicitly evaluative weights' must be established to assess and create policies that would achieve quality of life improvements, but that these criteria must be established in a context of 'open public discussion and critical scrutiny.' [69]

By his own admission, Sen is in these and other instances (e.g. Dreze and Sen 1989) developing a heuristic, whose utility may be defined primarily in terms of its ability to understand and represent complex social phenomena, such as poverty, vulnerability and freedom. But he is also quite clearly developing a normative position about the ways in which we (the academy, the industry, people) should act in relation to others, and in relation to universal norms of responsibility and action. Sen's idea of 'social commitment,' for instance, suggests a powerful sense of empathy, and responsibility:

> As competent human beings, we cannot shirk the task of judging how things are and what needs to be done. As reflective creatures, we have the ability to contemplate the lives of others. Our sense of responsibility need not relate only to the afflictions that our own behaviour may have caused (though that can be very important as well), but can also relate more generally to the miseries we see around us and that lie within our power to help remedy. That responsibility is not, of course, the only consideration that can claim our attention, but to deny the relevance of that general claim would be to miss something central about our social existence. It is not so much a matter of having exact rules about how precisely we ought to behave, as of recognizing the relevance of our shared humanity in making the choices we face.
>
> (Sen 1999 [2001: 283])

Three observations (among many) may be made about this particularly revealing quote. First, Sen is quite clearly calling for an ethic that goes beyond the calculation (or post-hoc rationalization) of narrowly-defined self-interest. Second, the principal source of morality – and of change – is the rational calculation of 'competent human beings' who can 'see' the world around them and 'contemplate the lives of others.' Finally, information, empathy and argument, it is assumed, may influence the decisions we make about the lives of others, and about our own norms, values and choices – a point carried forward in the following passage:

> Responsible adults must be in charge of their own well-being; it is for them to decide how to use their capabilities. But the capabilities that a person does actually have (and not merely theoretically enjoys) depend on the nature of social arrangements, which can be crucial for individual freedoms. And there the state and the society cannot escape social responsibility.
>
> (Sen 1999 [2001: 288])

An ethical choice is therefore unambiguously a matter of individual responsibility, which is shaped and constrained by the personal, social and environmental conditions under which people lead their lives. However, on what basis may rational, intelligent and fully informed people be expected to act in accordance with this theory of ethics?

Sen's morality suggests a fine line between a conservative liberalism, in which it is incumbent upon individuals (and, under certain conditions, states and societies) to take responsibility for their own actions; and a social commitment to social welfare, which requires that individuals, states and societies also take responsibility for the well-being of others. (We can surmise that where individuals lack the capability to exercise judgment or choice it is incumbent upon the state and society to act on their behalf.)

However, establishing whether people are able or unable to exercise rational judgment and choice raises myriad normative and methodological questions, which Sen is only partly able to resolve. First, the assumption that substantive freedom implies literacy, numeracy, freedom from avoidable disease, etc. is itself a normative statement about what constitutes 'the good life.' Although Sen's treatment of disability certainly broadens the neo-classical understanding of utility, and incorporates the ability of individuals (of different capabilities) to value and take advantage of different capability sets,[70] capability and disability are characterized almost exclusively in relation to one's ability to produce and consume, leaving little, if any, room for the normative terms on which 'disability' and difference may be valued or understood. The problem entails not (only) a question of respect for other interpretations of capability and value (which is important), but also a question about the means by which (and the moral grounds upon which) value is determined within society (Putnam 2002).[71]

For Sen, the challenge of achieving social consensus among what could conceivably be a limitless number of interpersonal preferences stems from three 'lines of scepticism,' all of which derive from neo-classical theories of individual decision making and rational choice. The first is Kenneth Arrow's 'impossibility theorem' that a voting system organized on the basis of preferential ranking will yield deadlock whenever there are three or more options (Arrow, cited in Sen, 1999 [2001]). The second is the assumption that any effort to discuss or design future scenarios will be plagued by uncertainty and indeterminacy; in Sen's words, 'things often do not go as we plan' (Sen 1999 [2001: 254]). Finally, there is the assumed problem that individuals are inherently self-interested, and that they will therefore reject any social compromise that contradicts their personal preferences.

To all of these dilemmas, Sen responds that public discourse and social institutions may provide a viable solution. To Arrow's impossibility theorem, he suggests that 'richer' information (about preferences, intentions, etc.) may change the win-set calculations of otherwise uncooperative agents. Echoing new institutional theories of uncertainty, rationality and collective action (see Chapter Two), Sen suggests that communication – and rules that would facilitate the transmission of information – may transform the possibility of social consensus on contentious issues. To the problem of unintended consequences, he makes the case that unintended consequences 'need not be unpredictable,' suggesting that causal analysis (and planning) may help to clarify and predict the possible range of outcomes that would follow particular decisions. Again the assertion being made is that institutional arrangements (governing the transmission of information, the calculation of risk, etc.) may help to refine the possible range of future outcomes. Finally, in response to the argument of self-interest, Sen argues that markets are functionally and historically dependent upon the existence of shared norms and values, reiterating the notion that the individual calculations and choices that people make in their lives are at least partly governed by wider sets of norms, values and historical social processes.

Therefore, Sen's conceptualization of social choice leaves open the possibility that communication, democratic institutions *and politics* may forge the conditions under which consensus about (potentially competing) interpersonal values may be achieved. However, the dilemmas he describes – and his solutions to these dilemmas – suggest a model that remains ambiguously rooted in a model of methodological individualism and rational choice. Theorizing the conditions under which shared norms and values may 'emerge,' for instance, Sen (1999 [2001: 273–4]) argues that social consensus may happen in a context of interpersonal 'reflection and analysis,' (individual) conformity to convention, public discussion and a (more weakly defended) process of 'evolutionary selection,' in which norms 'survive and flourish' on the basis of the 'consequential role' they play in governing social relations.

Second, the assumption that an expansion of choice would correspondingly lead to an expansion of capability presupposes that an expansion of freedom would (or should) have universal value, and avoids the classic liberal paradox of accommodating (or at the very least recognizing) the individual who chooses to enter slavery. Implicit in Sen's treatment of choice is the notion that freedom exercised individually or collectively (through social choice) would (or should) lead to an expansion of freedom to choose the life one wants to lead. However, as Gasper (2000) has argued, the freedom to choose may also 'come at the cost of much human freedom,' producing any number of undesirable 'side effects,' such as addiction, transmission of infectious disease, environmental degradation, etc., but also allowing the possibility that individuals may indeed choose slavery, tyranny, and so on. The dilemma has at least two dimensions: first, individuals may not fully or accurately 'know' the true consequences of their actions; although we may be able to establish (through science, experiential learning, risk assessment, etc.) a predictable range of possible outcomes (as Sen argues), the means by which we establish this predictable range is certainly open to error, and dispute (cf. Beck, 1992; Forsyth, 2003); and second, even if we know and understand the possible consequences of our actions, we may not care.

Whether and how the freedom to choose would deter people from making choices that would (in theory and practice) do harm to themselves or to others is unclear. For this reason, Sen must propose an 'additional' theory of freedom and responsibility, which imposes limits on the extent to which individuals may exercise their freedom, and seemingly contradicts his larger proposition that development constitutes the expansion of substantive freedom. In other words, the 'freedom' Sen has in mind is not the freedom to do anything one wants, but an expansion of freedom to pursue the life one wants (cf. Gasper 2000), *providing it meets the normative principles governing social commitment and collective responsibility.*[72]

Finally and equally problematic is the notion that the freedom to choose may require the existence *or application* of violence, coercion and other forms of unfreedom (Gasper, 2000). Implicit in Sen's treatment of democracy and freedom is the notion that social policy and government action are the result either of an over-arching ideology (e.g. a commitment to free market capitalism; to theories of food availability decline, etc.) or of what Sen has called (variously) public action or social choice.[73] Relatively underdeveloped in this formulation is the notion that decisions, mobilization, ideology, etc., may also reflect the material and symbolic power of different groups and interests within society. In the following section we consider more carefully the political conditions under which societies would mobilize to establish rights of freedom and social choice.

Political action and the freedom to choose … what exactly?

That the freedom to choose may be achieved on the basis of popular mobilization is relatively uncontroversial. Decidedly more contentious is the notion that popular mobilization (in and of itself) would necessarily incline towards the expansion of freedom. Even more problematic is the notion that popular mobilization will occur easily or unproblematically in a free market economy. Sen's treatment of entitlement and famine, for instance, suggests quite strongly that the deprivation he encountered in colonial Bengal (Sen 1981) and in other contexts (Sen 1981, 1999 [2001]) reflects the ways in which free market (and other market) economies organize and commodify labour. However, any link between the interests that underlie these modes, and the impact they may have on the performance of government (e.g. in India and China) remains largely unquestioned (Gasper 2000).[74]

Take, for instance, Sen's treatment of globalization. In *Development as Freedom*, Sen (1999 [2001]) devotes a substantial volume of text (25 pages, according to the index) to the issue of 'globalization.' Much of this (e.g. pp. 39–40; 120–7) entails a largely theoretical discussion about the perils of unregulated trade, and the need to intervene when markets turn against the poor. Sen's treatment of the Asian currency crisis (1999 [2001: 184–6]) and of India's liberalization experience (1999 [2001: 127]), for instance, suggests the importance of broadening social opportunities and of protecting populations made vulnerable by neo-liberal reform. Calling upon governments to intervene (e.g. by providing cash transfers), Sen strongly supports policies that would govern and at times suspend the 'rules of the game' when people are unable to command, through market mechanisms, the necessities of life.[75] However, nowhere in the analysis does he seriously challenge the existence of liberalized trade or (more importantly) associate the liberalization agenda with the ability of governments to provide important public goods, such as healthcare, welfare and education.

Where he does consider the impact of international power (1999 [2001: 240–8]), Sen is largely concerned with the globalization of 'Western' values, and the resultant impact on local economies and 'native cultures.' Sen's concern here is twofold: one relates to the impact of foreign trade and investment on 'traditional' modes of production and technologies; the second relates to the impact of economic and social interaction on traditional norms and cultures. To the first, Sen responds that 'societies' (defined presumably along national lines) must be prepared to adapt to globalization, by investing in education and other forms of training. To the second, he suggests that the loss of cultural traditions is 'an issue of some seriousness, but it is up to the society to determine what, if anything, it wants to do to preserve old forms of living, perhaps at significant cost' (Sen 1999 [2001: 241]). Beyond adaptation and survival, Sen concludes, little can be (or should be) done to resist the globalization of economic production:

> The one solution that is not available is that of stopping globalization of trade and economies, since the forces of economic exchange and division of labor are hard to resist in a competitive world fueled by massive technological evolution that gives modern technology an economically competitive edge.
>
> (Sen 1999 [2001: 240])

Whether and how this inevitability of globalization squares with Sen's understanding of freedom and of social choice becomes at this point rather unclear. In one sense, Sen seems to be suggesting that the freedom to resist globalization is an option, but one that would yield little (if anything) in the way of constructive dialogue. In another, he reiterates his faith in the value of public discourse. But the discourse he has in mind is one that requires 'societies' and governments to decide – through public discussion – the ways in which they will adapt (if at all) to the impact of globalization. The notion that certain groups and interests may be disproportionately able and/or inclined to support one policy decision (for instance favouring or resisting neo-liberal reform) over another receives no credible scrutiny.[76]

To all of this, one might respond (as does Hilary Putnam [2002]) that it was never Sen's intention to develop a theory of action. As Sen (1999 [2001]) argues in *Development as Freedom*, his aim is not to establish a precise formula that would assign a weight for every conceivable value, but instead to theorize the normative terms upon which evaluation (and therefore development) may be defined and pursued in different political, economic and cultural contexts (Putnam 2002). That having been said, theorizing the normative terms on which evaluation may happen must at some point take into consideration the practical implications that such a philosophy would entail.[77] Although he makes explicit the kinds of freedom he values,[78] Sen leaves ambiguous the kinds of power that would need to be in place to ensure that these and other substantive freedoms are recognized and preserved.

Sen's ambiguity about the nature and theoretical importance of power is important, not least because it leaves open to interpretation the kinds of power and interest that may facilitate the types of politics he has in mind, and also because his work has been so influential. From a methodological standpoint, the challenge of understanding what is desirable or what is important to (potentially) each and every individual involves the ability to communicate 'inter-subjective meanings' about the nature and meaning of value (over and around myriad individual worldviews of culture, ideology, etc.).[79]

The following section explores two bodies of theory that emerged in parallel and in response to Sen's work on capabilities and entitlement: participatory rural appraisal (PRA) and the sustainable livelihoods approach (the SLA). Set in a context of development management, both of these innovations aimed to emulate Sen's efforts to marry structure and agency and to develop a more open-ended approach to the study of poverty. Like Sen, the PRA and SLA's take was also highly ambiguous about the nature and impact of historical forces and social change.

Participatory approaches: from 'PRA' to the 'SLA'

Influenced in part by the recognition that standardized poverty assessments (i.e. income and expenditure surveys) offer an imperfect means of communicating complex concepts (like pain, suffering, family history, etc.), participatory theory advances the idea that a more accurate understanding of development will entail an effort to negotiate the ontological and metaphysical assumptions of poverty and of the research encounter. Rooted in liberalism, it also advances the idea that individuals (as well as individual groups and communities) have the ability to articulate and ideally realize their own views and preferences. By rejecting the reductionism and determinism of neo-classical theory, participatory theory aims to capture the complexity and nuance of poverty, identity and other aspects of development and, in so doing, inform better theory and practice.

Perhaps the most influential figure in this regard is Robert Chambers (1983, 1997), whose field-based empiricism aimed to provide a practical alternative to the heavy deductivism of orthodox Marxism, the artificiality of positivist social science and the apparent limitations of central planning and administration.[80] At the heart of Chambers' early work was the notion that development had become polarized between 'two cultures: a negative academic culture, mainly of social scientists, engaged in unhurried analysis and criticism; and a more positive culture of practitioners, engaged in time bounded action' (Chambers 1983: 28). The trouble for Chambers was that whereas the academic culture was often too critical, too distant and *too slow* to have an immediate or positive effect on the vast majority of the world's poor, the practical culture was too deeply constrained by organizational imperatives and too uninformed about historical context and ground realities (cf. Edwards 1993).[81] The challenge, he argued, was to combine the analytic and experiential merits of the academic and practical cultures, and to incorporate a third: that of 'the rural people in a particular place' (Chambers 1983: 46). In this way, Chambers advocated what he felt was a paradigmatic shift away from the stifling 'pessimism' of historical materialism and the condescending paternalism of central planning and neo-classical economics towards a more participatory and inclusive understanding of development and practice.

Working from traditions of farming systems and agricultural extension, Chambers (1983, 1997) shares a number of assumptions with Sen about the limitations of conventional poverty assessments, and the related need for more participatory forms of inquiry. At the heart of Chambers' participatory approach is the idea that the gap between the framed assumptions of researchers/donors and the 'real' needs of those in need may be overcome by creating the conditions under which the objects of development research and/or of development would more actively contribute to and ideally define the aims and ends of development and of development research.

Chambers' principal strategy – and one that has since been widely adopted and emulated – is a practice he termed 'Participatory Rural Appraisal' or PRA. At the heart of PRA is the idea that, by changing the assumptions and practice of development research and development projects, development researchers and professionals can create for poor people (whose definition was open to interpretation) a more meaningful opportunity to participate in the definition and direction of development research.[82] Towards this end, he promoted a wide range of strategies that would (ideally) remove or reduce: (1) the power imbalances between the authors and objects of social science research; and (2) the biases of academic, urban, cosmopolitan, 'modern' experience that researchers bring to the research process. Included here are assessments of poverty and well-being that replaced or supplemented:

- pre-determined survey questionnaires with participatory ranking exercises, involving stick diagrams, social mapping, etc.;
- standard household surveys with participatory observation, in which researchers actively participate in the labour and other livelihood activities of the poor;[83]
- isolated and artificial elite interviews with focus group discussions, which leave open the potential range of issues, and (ideally) give individual respondents the opportunity to make themselves heard.[84]

Distancing himself from Marxism, dependency and other forms of historical materialism, Chambers (1983: Chapter 5) contends that poverty may be defined (and explained) in terms of 'clusters of disadvantage.' Like Sen, Chambers suggests that poverty is reflective of the assets that households and individuals have at their disposal, as well as the personal, environmental and political factors that affect their ability to produce and consume in a free market economy. What makes individuals and households poor and/or susceptible to poverty includes *structural factors* (e.g. isolation from schools, markets and hospitals; chronic illness; dependency ratios; vulnerability to sickness and disaster; powerlessness) and *idiosyncratic factors* – what he calls 'poverty ratchets'(Chambers 1983: 114) – that force individuals and families to deplete the few assets they have at their disposal. Included here are five types of contingency:

- social conventions (e.g. dowry inflation);
- natural disasters;
- physical incapacity, stemming from sickness, 'the child-bearing sequence of pregnancy, childbirth, and the post-natal period,' and accidents;
- unproductive investment; and
- exploitation.

Like Sen, Chambers therefore incorporates questions of structure, contingency and power to explain the factors that may lead to deeper and increasingly irreversible forms of poverty; poverty and destitution are defined in a way that incorporates the assets that individuals and households have at their disposal, as well as the personal, environmental and political factors that may force them into poverty spirals. Further refinement of the PRA methodology would lead to the 'sustainable livelihoods approach' (SLA), which is a research and planning tool that became very popular among development researchers (especially ones funded by or working for the UK DFID).[85]

Creating capabilities: the 'sustainable livelihoods approach'

Drawing upon the work of Chambers and Sen, the SLA embraced the importance of structural vulnerability, household decision making and the micro-economic impact of macro-economic processes and 'idiosyncratic' events. The following quote from Chambers and Conway (1991) provides the conceptual basis for subsequent work:

> A livelihood comprises the capabilities, assets (including both material and social resources) and activities required for a means of living. A livelihood is sustainable when it can cope with and recover from stresses and shocks, maintain or enhance its capabilities and assets, while not undermining the natural resource base.
>
> (Chambers and Conway 1991; paraphrased by Scoones 1998: 5)

Central to the SLA is the notion that individuals possess 'a portfolio' of assets, which are fungible in nature and subject to the logic of investment. Strongly influenced by neo-classical assumptions about household decision making, poverty and well-being are associated with *five types of capital asset*:

- *Natural capital*: the physical inputs that generate value and productivity in people's lives. Examples here would include rainfall, soil, sunlight and climate;
- *Human capital*: the skills, knowledge and labour that people use to make a living;
- *Financial capital*: the sum total of one's material possessions and investments;
- *Physical capital*: the basic infrastructure and means by which people engage in economic activity;
- *Social capital*: the networks of trust, reciprocity and loyalty that enable people to thrive and survive in economic life.

Framed in this way, poverty is 'strongly associated with a lack of assets or an inability to put assets to productive use' (Ellis 1998: 59). Capital assets are productive in the sense that they facilitate ends that would not be attainable in their absence. For instance, financial capital can be used to acquire other forms of capital, including human capital (labour and knowledge), social capital (loyalty, networks) and natural capital (water, land). Likewise, the theory goes, human capital can be used to acquire economic, social and natural capital, and so on. An important point here is that one form of capital can be used to acquire other forms of capital. Therefore the ability to cope with and ideally 'escape' from poverty entails an ability to invest one's assets in a way that is productive and sustainable.

In terms of methodology, the SLA embraces many of the strategies pioneered by Chambers and by PRA. Included here are participatory methods, such as participant

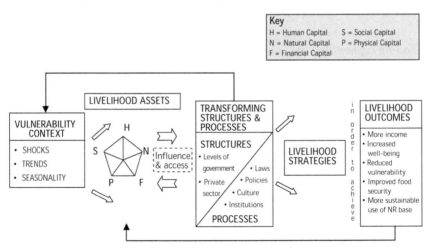

Figure 5.1 DFID's Livelihoods Framework.

Source: DFID, 1999.

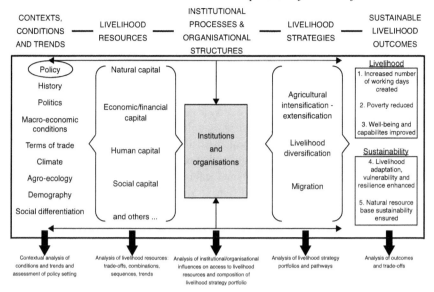

Figure 5.2 Sustainable Livelihoods (Scoones, 1998).

Source: Scoones, 1998.

observation, key informant interviews, focus group discussions (FGDs) as well as household surveys, which are used to understand the ways in which different groups perceive and respond to myriad forms of poverty and risk. Building upon Chambers and on PRA, the SLA also theorizes in much greater detail the role of context, institutions and macro-systemic 'trends' in the decisions that individuals make about livelihoods and risk. Under the heading of 'policy,' for instance, Ian Scoones (1998) includes politics, history, macro-economic conditions, terms of trade and other macro-systemic variables whose individual and combined impact could in theory affect the well-being of the poor. Along similar lines, Susanna Davies (1996) distinguishes between positive and negative adaptation strategies. 'Positive adaptation,' she argues,

> is by choice, can be reversed if fortunes change, and usually leads to increased security and sometimes wealth. It is concerned with risk reduction and likely to be an intensification of existing livelihood strategies; or diversification into neighbouring livelihood systems.

(Davies 1996: 5)

'Negative adaptation' is of necessity, tends to be irreversible and frequently fails to contribute to a lasting reduction in vulnerability. Negative adaptation occurs when the poor are forced to adapt their livelihoods because they can no longer cope with short-term shocks and need to fundamentally alter the ways in which they subsist.

(Davies 1996: 5)

Another innovation concerns the role that 'policies, institutions and political processes' (PIPs) may have on the lives of very poor people. In Scoones (1998) and (especially) in DFID (1999), large emphasis is placed on the importance of 'structure' (including laws, 'cultures,' policies and rights) and of politics, framed primarily in terms of individuals and individual households using their assets (e.g. legal literacy, social capital) to improve access to a wide range of economic and social opportunities. In so doing, the SLA suggests that poverty is not only the result of micro-economic decisions and macro-economic change, but also of the services that governments may provide during times of economic hardship and stress. (See Box 5.2)

Underlying the SLA is an assumption that participatory research methodologies may yield more accurate insights about local realities, which would in turn inform better policy. Like Chambers and Sen, it aims to use systemic/structural and idiosyncratic factors to understand the ways in which individuals and households cope and adapt during periods of uncertainty and stress (Bebbington 1999; Scoones 1998). Next, I assess the ability of PRA and the SLA to generate theory about the nature of poverty, livelihoods and social change.

Assessing the SLA

The perceived value of the SLA rests in its ability to provide a more flexible means by which researchers and donors could investigate and understand poverty and development. First, it embraces the assumption that poverty and well-being are dependent upon structural forces (e.g. terms of trade, demographics, agro-ecology, etc.), but also upon the ways in which people 'invest' their assets and resources in productive activities. Second, it recognizes that people's livelihoods may vary with seasons and with prevailing economic conditions ('shocks and stresses'), and that poor people may be engaged in a variety of livelihood activities, often at the same time. Third, it suggests that efforts to understand and affect poverty require new ways of thinking outside of 'traditional' academic and sectoral categories (e.g. labour, capital, the peasantry; health, education, agriculture) to support the multiple and diverse livelihoods of the poor.

From a project management perspective, the SLA provides an attractive means by which programmers may assess poverty and, on the basis of this assessment, design projects that would more accurately reflect the complexity of local needs and context (see Box 5.2). As DFID's 'Sustainable Livelihoods Guidance Sheet' suggests, the livelihoods framework offers 'a way of putting people at the centre of development, thereby increasing the effectiveness of development assistance' (DFID 1999: 1.1):

> Like all frameworks, it is a simplification; the full diversity and richness of livelihoods can be understood only by qualitative and participatory analysis at a local level ... The framework does not provide an exact representation of reality. It does, however, endeavour to provide a way of thinking about the livelihoods of poor people that will stimulate debate and reflection, thereby improving performance in poverty reduction.

(DFID 1999: 1.1)

In retrospect, PRA and the SLA provide a healthy alternative to the world of development professionalism – a world that Chambers, Scoones, Davies and Conway know very well. For many development agencies (such as DFID, UNICEF, CIDA and UNDP), the strategy provides an attractive means by which development professionals can develop and refine the goals and practices of development projects, and assess the real and perceived impact of their interventions in the field. However, the notion that participatory approaches will yield a more accurate or authentic understanding of reality and of development is deeply problematic in a number of ways. First, PRA and the SLA are at times naively optimistic about the ability of participatory approaches to overcome the inequalities of power that structure 'local' society (Kapoor 2002; Mohan and Stokke 2000; Mosse 2001, 2005; Kothari 2001). Subject to considerable scrutiny in the literature that has emerged in critical response to PRA is the idea that 'egalitarian' approaches to rural development research may overcome or address longstanding structures and inequalities of power (e.g. Kapoor 2002; Kothari 2001; Hickey and Mohan 2005). For instance, Ilan Kapoor (2002) questions the ability of PRA to induce,

Recent debates in the development policy literature have popularized the idea of using cash transfers to protect vulnerable groups from the loss or reduction of incomes during periods of market instability. The idea here is that the inability to meet basic needs (of nutrition, healthcare, etc.) during droughts, famines and other volatile periods is not generally the result of insufficient food or medicine, but the result of insufficient buying power on the part of affected populations (Sen 1999 [2001]; Farrington and Slater 2006). Therefore, the perceived solution is to provide affected populations with cash transfers, which in theory would improve their ability to command scarce resources in free market economies (Dreze and Sen 1989). While there is general consensus that cash transfer programmes can help to reduce or relieve poverty, scholars differ about the means by which such aims can and should be achieved. Barrett *et al.* (2004), for instance, argue that public works programs should lead to the provision of public goods such as infrastructure, which can then have 'second round' effects on local economic development. Others believe that cash transfer programmes should focus on the primary objective of creating employment to meet the more immediate needs of the very poor and vulnerable (Dreze and Sen 1989). However, a problem that exists in many low-income economies is that employment creation programmes are notoriously difficult to administer, raising questions about the impact that governance can have on the livelihoods of the poor (Johnson *et al.* 2005; Deshingkar *et al.* 2005; Dreze 2006).

Box 5.2 Supporting sustainable livelihoods: cash transfers through lived observation, experience and practice, insights about the nature of poverty, vulnerability, etc., suggesting that PRA is theoretically ill-equipped to deal with questions of power and difference.[86] Others (such as Hickey and Mohan 2005; Kothari 2001; and Mohan and Stokke 2000) raise the concern that poverty, vulnerability, etc., are a product of wider historical forces and transformations that PRA (and, by extension, the SLA) is poorly equipped to comprehend or explain.

Second, the idea that participatory approaches may capture more accurately or authentically local needs and realities underplays significantly the impact that PRA (or, for that matter, any form of bureaucratic practice) may have on the perceptions and behaviour of its intended beneficiaries (Mosse 2001; Kapoor 2002). David Mosse (2001), for instance, has shown with great nuance the ways in which villagers in western India were able to re-define (or in Foucauldian terms 'normalize') themselves in the context of participatory research. Knowing full well the norms and expectations attached by donors and project implementers to 'the participatory experience,' the villagers Mosse encountered in his research quite convincingly emulated and negotiated the norms the project was hoping to see, suggesting that the accuracy and authenticity of participatory projects may be affected by local efforts to do and say the things that donors and researchers want to see and hear.

Third, the idea that participatory approaches may represent the 'real' needs and perceptions of the poor is, in practice, structurally constrained by the fact that it has been most significantly promoted, embraced and applied in a context of development management (Kapoor 2002; Parfitt 2004). As Parfitt (2004) has argued, Chambers' argument in favour of PRA is tied explicitly to the notion that 'better' participation would improve the efficiency and effectiveness of development projects and policy. When applied by large development agencies, he argues, the result is often

> ... a mechanistic, routinised, largely symbolic practice of participation (often in the form of a rather perfunctory PRA) co-existing with an operational policy of traditional top-down management that affords the recipients little in the way of empowerment.
>
> (Parfitt 2004: 547)

Similar arguments have been advanced by Kapoor (2002), Kothari (2001) and Mosse (2001, 2005).

Fourth, the idea that participatory research could contribute to a body of empiricism that would lead to higher forms of abstraction underplays significantly the more general lack of theoretical reflection about the relationship between lived experience and the construction of knowledge (Kapoor 2002). Like any good case study, PRA and the SLA are most convincing when they test, contradict, expand or refute pre-conceived notions (or hypotheses) about the nature of poverty, development, farming, and so on. However, the ability of PRA and the SLA to generate insights that go beyond the field site is constrained significantly by their inability to reflect (either historically or theoretically) on the connection between the data generated by means of participatory research methods and wider bodies of knowledge and practice.[87]

Finally, the idea that livelihoods and poverty may be measured and assessed primarily in relation to a 'portfolio' of assets used by individuals (and individual households) to cope with the various stresses and shocks of a market economy conjures an image that is squarely rooted in the neo-classical frame. For instance, Ben Fine (2001) has argued that the construction and consolidation of 'social capital' within the World Bank entails a radical retreat from the structuralist/ historicist theorizing of Pierre Bourdieu (who pioneered the term) and favours instead what is essentially a neo-classical rendering of the work of James Coleman (1990) and Robert Putnam (1993) (cf. Putzel 1997; Harriss and de Renzio 1997; Harriss 2001). At the heart of this rendering is an image of society and of reality governed by individual decision making and rational choice.

Set in the context of development administration and project management, PRA and the SLA therefore advance a methodology that aims to collect information that has any conceivable connection to people's livelihoods or to the sustainable livelihoods framework. The result is an approach (to project management and to research) that is very good at providing 'snapshots' of rural poverty, suffering, deprivation, etc., but fails to establish explicitly a meaningful connection between the methodology being promoted and the historical conditions under which the research and the object of the research (i.e. poverty and deprivation) occurs.

Concluding remarks

Our central aim in this chapter was to consider the extent to which the innovations developed by Sen, Chambers and the SLA offer a theoretical and methodological alternative to the neo-classical orthodoxy. Its central claim is that Sen's re-conceptualization of poverty, capabilities and entitlement helped to advance development theory in a way that went beyond the neo-classical model of individual decision making, marginal utility and rational choice. Where Sen is more ambiguous and more firmly wedded to the neo-classical frame is in relation to his explanation of the factors that would lead to an expansion of freedom and choice. As the preceding suggests, Sen is at times highly ambiguous about the empirical and normative conditions under which basic rights of freedom and entitlement would be respected and accommodated within society. On the question of social consensus, for instance, he suggests that common norms and values may be established on the basis of inter-subjective communication and the institutionalization of democratic practices and values. On the question of globalization, Sen's analysis is at times highly ambiguous about the viability and desirability of resisting the forces of neo-liberal reform. The best that societies can do, Sen suggests, is to adapt to globalization by developing institutions that would protect populations made vulnerable by liberalized trade. As realistic as this may be, it is difficult to reconcile the inevitability of globalization with the notion that 'good societies' are ones that allow individuals the freedom to choose, through open and democratic discourse, the lives they would want to lead.

Reflecting upon the intellectual impact of Sen's lifetime body of work, Des Gasper (2000) has argued that Sen's ability to appeal to a broad range of intellectual and ideological perspectives reflects both the interdisciplinarity of his writing and his ability to 'speak the language' of neo-classical economics. Although he clearly distances himself from the neo-classical frame, Sen's defence of individual liberties and the freedom to choose at times slips into a logic of individual decision making and rational choice, highlighting the ideological and methodological importance of adopting or at least speaking the language of neo-classical discourse.

A second insight we can draw from the preceding chapter is the perceived value of 'opening' the terms of social science inquiry about the empirical nature of economic development and social change. As we have seen, PRA and the SLA share an assumption with Sen that poverty and deprivation are the combined result of 'systemic' and 'idiosyncratic' factors, which structure the choices individuals and individual households can make about the ways in which they employ cash, labour and other household assets. Robert Chambers (1983), for instance, describes the ways in which illness, dowry, death and systemic factors (e.g. inflation) may 'ratchet' poor people further into poverty. Similarly, Scoones (1998), Davies (1996) and the SLA suggest that poverty and deprivation are the result of systemic factors, individual assets and idiosyncratic 'shocks and stresses' that may push people even further into poverty.

Compared with the 'heavily deductive' approaches of dependency and Marxism, the participatory approaches of PRA, the SLA and Sen's capabilities approach give much greater attention to the finer details of household decision making and life cycle needs and events. In so doing, they provide a powerful means of capturing the complexity and diversity of (primarily local) development processes, generating insights about micro-level impacts of macro-economic policies and change. However, the notion that participatory methodologies may provide the basis upon which universal explanations could be based became a matter of substantial debate. First, the notion that participatory research strategies would yield a more accurate understanding of local realities has been criticized for placing excessive explanatory weight on local factors and on the statements of individual respondents (Mohan and Stokke 2000; Mosse 2001, 2005; Kapoor 2002). Second, participatory methods have been criticized for ignoring or under-stating the historical structures and forces that shape the perceptions and statements made by individuals during participatory research (Kapoor 2002; Kothari 2001). Third, they are at times naively uncritical about the power they imposed *or displaced* during participatory research exercises (Kapoor 2002; Mosse 2001, 2005). Finally, they fail to connect their methodology and their empiricism with larger historical systems and forces, and with the construction of theory (Leys 1996; Kapoor 2002; Mohan and Stokke 2000).

To these criticisms, it may be argued that the participatory paradigm was never intended to provide a formal methodology. Indeed the vast majority of writing about PRA and the SLA has been directed primarily towards an audience of development professionals and agencies, whose principal aim is to develop more effective programs and policy, not to advance a theory of knowledge. As Trevor Parfitt (2004) has argued, the importance of participatory research relates not only to the contribution it makes to a predictive social science, but also to the opportunities it provides for open discourse about the means and end of development.

The idea that participation is only or primarily about developing methods through which social scientists (and development professionals) may frame and undertake better field research of course dramatically underplays the idea that participation may also entail a more substantive agenda aimed at expanding the political terms of discourse, citizenship and empowerment (Sen 1999 [2001]; Parfitt 2004; Hickey and Mohan 2005). Again, to what extent and on what basis may these 'transformative' agendas be theorized, idealized and pursued?

The following chapter explores the theoretical and political challenges that an agenda of this kind would entail.

6 Advancing knowledge for social change

> Without a historically deep and geographically broad analysis, one that takes into account political economy, we risk seeing only the residue of meaning. We see the puddles, perhaps, but not the rainstorms, and certainly not the gathering thunderclouds.
>
> Paul Farmer (2004: 309)

> Modern history begins when history becomes concerned with the future as well as with the past.
>
> E. H. Carr (1951: 2)

Introduction

The preceding chapters have described a field that has become increasingly inclined towards the rejection of grand ideas about the nature of development, the course of history and the need for social change. Instead of revolutions and grand social forces, the dominant voice in development is now one that invokes goals and targets to reduce poverty or – more cynically – to justify spending increases in foreign aid (Mosse 2005; Eyben 2006; Saith 2006). As noted in Chapter One, the problem is not that the MDGs are unworthy in their own right. Rather, it is that they are framed in a language that fails to articulate with sufficient clarity or purpose the ways in which they would address the structural factors and historical processes that perpetuate the conditions (poverty, exploitation, gender inequality, environmental degradation, etc.) they aim to address.

In this book I have argued that the retreat from grand social theory (and politics) in development reflects a number of historical factors. One has been the de-legitimation of historical social analysis in positivist (and especially American) social science. A second is the internal fragmentation of the Marxist paradigm. A third is the rise of an outlook that questions the ability of history and social science to represent other histories, cultures and people. A fourth and more general factor is the now dominant role of neo-classical theory in development theory and research.

Although it offers an important means by which scholars may expose and deconstruct the discourses of large development agencies, such as the IMF and the World Bank, post-development is in many ways a poor substitute, offering visions of the future that either engage in unhelpful rhetoric about the underlying nature of capitalism, modernization and 'the West' or ones that obviate themselves of any apparent need to connect theory and praxis. Even Sen, who provides possibly the most important and coherent response to neo-liberalism, slips into a logic of individual decision making and rational choice, highlighting the ideological and methodological importance of adopting or at least speaking the language of neo-classical discourse.

This chapter concludes the book by situating these trends in a wider historical context and outlining a number of ways in which development studies may contribute to a body of knowledge that is inclusive, rigorous and engaged. To illustrate what I believe to be at stake in this analysis, I first want to bring us back to an analogy I used at the beginning of the book: development research, development policy and the World Bank. Next I outline two ways in which development theory may recapture the kinds of 'grand' social theory and analysis envisioned by Tilly (1984), Evans (1995) and, more recently, by Kohli (2004) and Sandbrook *et al.* (2007). The final section concludes the book by reflecting on the political and philosophical limitations of connecting social theory with a broader ideology of social change.

'Making services work for the poor'

In 2004, the World Bank (2004) unveiled an ambitious plan to improve the delivery of basic services – such as primary healthcare, universal education and clean drinking water. In its *2004 World Development Report* (World Bank 2004) the Bank made the case that the delivery of public services can improve if poor people are given the opportunity to monitor and enforce the behaviour of public officials. Alongside measures that would improve the transparency of government, amplify the 'voices' of poor people and punish/reward inappropriate behaviour on the part of public officials, improvements in public services can best be achieved by expanding the choice that 'poor clients' have in the selection of public services.

To support its case, the Bank (World Bank 2004) makes an important distinction between the 'long route' to accountability, in which 'clients' must 'go through' policymakers in order to influence and affect the behaviour of service providers, and the 'short route,' in which the transaction – and therefore accountability – implies an immediate and direct relationship between the citizen/client and the service provider. Citing school voucher programmes in Colombia and (partial) pay for service drug programmes in El Salvador and Guinea, the report suggests that subsidized or private payments on the part of the poor can 'enable clients to exert influence over providers through choice' (World Bank, 2004: 6).[88] Framed in this way, citizenship – defined on the basis of monetary exchange – provides the formal basis upon which stronger claims for effective and accountable service delivery can be made.

The theoretical core that underlies the World Bank's suggestions about what makes for better service delivery is an old one. Put simply, it suggests that the right to demand satisfaction from government – or from any service provider – is rooted in the transfer of individual liberties, especially in private property. Prior to the establishment of liberal democracy in Western Europe, this 'transfer' was typically *non-voluntary* in the sense that labour and the fruits of one's labour were often 'extracted' without the consent of the provider (see especially, Tilly 1990: Chapter 3; Levi 1988). However, over time, and on the basis of many different social transformations, there emerged a normative and procedural link between the non-voluntary extraction of individual liberties (most commonly in the form of taxation) and the demand that government provide certain goods and services in ways that conform to historical norms and expectations (Ignatieff 2000; Macpherson 1973; Redden 2002; Tilly 1990).

By encouraging arrangements in which clients pay for services – either out of their own pockets or (in the case of voucher schemes) out of the pockets of government, donors or others – the World Bank (2004) is suggesting that the relationship between clients and service providers can be made more direct and therefore accountable. Moreover, and unlike the European model, it is suggesting that the transfer can be made *voluntarily* through the provision of choice. The idea here is, firstly, that 'choice' will give clients the power to 'take their business elsewhere,' so to speak, thereby encouraging service providers to improve their delivery (Brett 2000); and second, that the assumption of cost on the part of the client will create an incentive to monitor the performance of the service provider (World Bank 2004).

Reforming traditional modes of public administration (by encouraging decentralization) and public service delivery (by introducing vouchers, charter schools, privatization, etc.) can re-structure incentives in a way that improves the ability of government to produce (or 'co-produce') important public goods (Stein 2001; World Bank 2004). However, the assumption that citizens will demand better services and participate in democratic institutions as do clients in a market exchange, adopts a model of citizenship and markets whose roots are firmly anchored in the Western liberal democratic tradition of private property, modern taxation and public administration.

In Western democracies, debates about accountability and public services have become increasingly dependent upon the idea that the relationship between citizens and government can *and should* be understood as a relationship between clients and service providers. Writing about citizenship and entitlement to healthcare in Canada, for instance, Redden (2002) has argued that the rise of the client culture is part of a wider shift away from traditional notions of citizenship in which collective responsibility and entitlement were intimately connected, to a more atomized understanding in which the individual transfer of property constitutes the fundamental basis upon which modern citizenship is based. Along similar lines, Stein has argued that,

> The culture of choice, part of the larger tapestry of radical individualism, is nourished by the sense that government is insufficiently responsive, and that we as citizens are quite capable of making sound judgments on public issues.
> (Stein 2001: 82)[89]

For societies with large populations employed in informal sectors of the economy, linking the provision of services on the part of government and the extraction of revenue on the part of citizens is, in a number of ways, deeply problematic. First, 'revenue collection' in low-income countries tends to be highly informal (e.g. through tariffs, fees, bribes, etc.) and indirect (through national tariffs, the suppression of farm gate prices, etc.). The ability to articulate and demand entitlement on the basis of rights is therefore deeply constrained by the lack of direct connection between 'taxation' and government service. Moreover, the costs of collective political action (e.g. costs of travel, communication and/or potential backlash) may deter poor people from demanding better service from government (Moore and Putzel 1999).

Second, the separation between public and private life is an imperfect one, fostered in large part by the enormous role (particularly in formerly import-substituting economies, such as India) that government has played in economic life and by the compromised nature of the post-colonial state (e.g. Migdal 1988). In many low-income countries, the incentives that motivate public officials tend to be skewed heavily in favour of individual gain, and generally at the expense of the poor (indeed, the very reason people go into public life in the first place is to achieve the formal authority, which would facilitate informal gain – see Wade's [1985] classic treatment).

Third, notwithstanding some form of subsidy, requirements that payments be made in exchange for services will be disproportionately costly for the poor.

Finally, the notion that accountability can and should be articulated on the basis of proprietary rights (i.e. voluntary and non-voluntary extractions of individual liberties) is deeply rooted in a Western liberal democratic tradition, and may not be consistent with the ways in which individual and collective rights have been articulated and defended in societies lacking this tradition. In India, for instance, one could make an equally persuasive case that government's authority to rule is rooted in a profound sense of injustice which, over time, has created an obligation on the part of the state and society in general to 'compensate' the groups and individuals whose suffering reflects a prolonged history of social injustice (see especially, Khilnani 2002; Saberwal 2002; Chandhoke 2002; Galanter 2002). Viewed in this way, *reservations*, which provide special treatment (e.g. quotas in government bodies, public education, employment programmes, access to housing, and so on) for 'scheduled' castes and tribes, and *social welfare policies* (such as food for work and employment guarantees), can be understood as an entitlement to which groups and/or individuals in society have legitimate claim, irrespective of their 'ability to pay.'

In short, the idea of applying a neo-classical model of accountability (e.g. World Bank, 2004) to societies in which Western traditions of taxation, representation and the rule of law cannot be assumed is strewn with problems of logic and application. Moreover, the notion that a neo-classical model of governance and accountability can or should replace pre-existing institutions governing social, political and economic life opens an ethical hornets' nest of questions about the rights of communities, individuals, etc. to establish or maintain their own institutions and traditions. At the risk of belabouring the point, such concerns appear to be of little interest to neo-classical theory.

Debating the discipline: big theories, local processes and the art of comparison

As noted in Chapter One, alternatives to neo-classical theory are often framed in terms of re-capturing the politics and history of development. Framed in this way, the challenge of *'re-politicizing'* development entails the (re)-establishment of an interdisciplinary paradigm that is problem-oriented, action-oriented and necessarily geared towards the construction of general theory about the historical forces that affect questions of distribution, deprivation and material well-being.

The preceding chapters describe a field that has become increasingly fragmented in terms of the theories, concepts and methodologies it uses to understand and explain complex and contextually specific processes of economic development and social change. Outside of neo-classical theory (and related fields of rational choice), the notion that social science can or should aim to develop general and predictive theories about development has become mired in a philosophical and political orientation that questions the ability of scholars to make universal or comparative statements about the nature of history, cultural diversity and progress. The result is a field that has become extremely good at documenting the nuance and complexity of local development processes, but rather less good at connecting these ground realities to wider, historical trends and forces.

Chapter One raised a number of concerns about the apparent lack of comparative methodology in development research. For one, the emphasis on local processes appears to be too far removed from the larger structural and historical transformations that now shape the political economy of development (cf. Peet and Hartwick 1999; Fine 2001; Hoogvelt 2001; Mohan and Stokke 2000). Second, the emphasis on local processes and case studies appears to limit the range of options that would in theory inform efforts to connect social theory and social change (Leys 1996; Mohan and Stokke 2000; Edwards 2002, 2006). Although ethnographic and case study research certainly have a role to play in the construction of knowledge (exploring for instance, causal mechanisms, testing or investigating the validity of theories and concepts, falsifying theoretical assumptions and producing new theories and hypotheses about social processes and events), an overriding concern is that development has become exceedingly dependent on the documentation and analysis of local and locally contingent processes and events.

Reflecting on the state of the art of development theory (circa 1990), Colin Leys (1996) argued that the 'grand' theories of Marxism, dependency and development had by the end of the 1980s been subsumed by a view of the world that explained political and social outcomes either in terms of individual decision making and rational choice or in terms of a 'discourse of "complexity,"'

... in which everything is dissolved into its details, and the possibility of abstracting and trying to act on the main elements and forces at work in the world is obscured (if not actually denied).

(Leys 1996: 196)

Ten years later, in a keynote speech to the 40th anniversary of the founding of the Institute of Development Studies at the University of Sussex, Michael Edwards lamented what he felt was an excess of disembodied empiricism, calling upon development researchers to develop 'more systematic efforts to "join the dots," to make the connections, to identify patterns of cause and effect across time and space, to place individual experiences in their wider context' (Edwards 2006: 5). Reflecting on the period in question, John Harriss suggests that:

Most of us were inclined towards 'micro' studies rather than towards macro, and though we often sought to analyse the linkages between societal levels ... too few of us examined trends in the global economy and their implications, except at the level of generality of much of the work in the dependency and world systems theoretic traditions.

(Harriss 2005: 30)

The argument being made here is not that development needs to do less ethnography. Rather, and crucially, it is that development needs to do more comparison. As Charles Tilly has argued, the value of doing 'huge comparisons of big structures and large processes,' is that they:

... help to establish what must be explained, attach the possible explanations to their context in time and space, and sometimes actually improve our understanding of those structures and processes.

(Tilly 1984: 145)

Taking on board the arguments that may be levelled against comparison (e.g. Wallerstein 1986; McMichael 2000) and generalization (e.g. Johnson 2006), an important challenge therefore is to devise a way of developing comparative social analysis that can incorporate the nuance of history and context, while at the same time providing the basis upon which inferences (and therefore actions) may be based. As Paul Pierson reminds us,

... the point is not that we need to know everything about the context of a particular phenomenon – which is not just a practical but a logical impossibility ... The point is that what is too easily dismissed as 'context' may in fact be absolutely crucial to understanding important social processes.

(Pierson 2004: 169)

On what basis and in what ways may development therefore become more comparative in nature and scale?

Implicit in the positions being taken by Booth (1985, 1993), Scott (1985), Chambers (1983) and Sen (1981; 1999 [2001]) is an assumption (indeed a faith) that research can be made receptive to questions of diversity and difference while at the same time productive in relation to the construction of theory. Booth (1993), for instance, suggests that the way forward is to identify and employ 'bridging themes' that could illuminate the micro-foundations of macro-processes, and also understand, through the use of local and case study research methods, issues that have clear and logical connections to wider systemic or historical processes.[90] Similarly, Vandergeest and Buttel feel it is 'possible to avoid the teleological assumptions of Marxist development sociology without lapsing into empiricism' (Vandergeest and Buttel 1988: 687).

The following sections consider two possible ways in which development research may engage in broader and more comparative forms of analysis. One, rooted squarely in the positivist tradition, aims to combine the rigor of theory and the nuance of history by using historical narratives to frame and test hypothetic–deductive models of social behaviour. A second provides an inductive means by which scholars may search for commonalities and connections to broader historical trends and problems, while at the same time incorporating divergent and potentially competing views about the nature of history, culture and context.

'Analytic narratives'

As noted in Chapter Two, positivism has traditionally eschewed the idea of ascribing causal power to historical trends and forces. However, more recent approaches have introduced the idea of combining the perceived rigor of formal modelling and statistical analysis with the contextual nuance and detail of historical narrative. Reflecting on the 'state of the art' in institutional scholarship, for instance, Campbell and Pederson (2001a) have described what they perceive to be a 'second movement' in institutional analysis, in which historical institutionalists and rational choice perspectives (along with the work of 'organizational' and 'discursive institutionalists') are merging to produce more productive forms of institutional analysis. Central to their case is the assertion that first, there is 'no a priori reason' that methodological differences (between historical and rational choice perspectives) should prevent the combination of hypothesis testing and 'thick' description and second, there are complementarities – of arguments, insights and problems – that can be developed to merge the two approaches (Campbell and Pederson 2001b).

Similarly, King *et al.* (1994) argue that science and cultural/historical interpretation 'are *not* fundamentally different endeavours aimed at divergent goals' (King *et al.* 1994: 37). On the contrary, they contend that historical and contextual narratives can complement scientific methods 'by helping to frame better questions for research' (King *et al.* 1994: 38). In this respect, their conclusions are very consistent with those put forth by Bates *et al.* (1998), in which 'analytical narratives' are advanced by combining historical analysis with formal modeling (developed largely in the tradition of game theory and rational choice).

Although one would not want to draw too fine a line of congruity,[91] current arguments in favour of complementary or 'tripartite' (Laitin 2003) ways of combining historical narrative, deductive modelling and the construction of general theory (e.g. Bates *et al.* 1998; Campbell and Pederson 2001a, 2001b; King *et al.* 1994; Laitin 2003) share with Popper (1957 [1997]) the idea that social science needs to inject a 'preconceived selective point of view into one's history' (Popper 1957 [1997]: 150), which would permit the testing of competing and falsifiable hypotheses.

Take, for instance, *Designing Social Inquiry* by Gary King, Robert Keohane and Stanley Verba (King *et al.* 1994).[92] Although the authors do not describe themselves as 'positivists' (Johnson 2006), King *et al.* (1994) share with positivism the assumption that social science should be established on a logic of inference, which establishes causal and/or descriptive regularities through empirical research, using 'public' (as opposed to private or 'esoteric') methods in which conclusions are uncertain, falsifiable, and contingent upon a recognized system of inference. Framed in this way, the formulation of questions, concepts, hypotheses and methodologies is carried out in isolation from the objects of social science research.

Central to their logic of inference is the idea that researchers develop proxies or 'observable implications' of the phenomena they want to measure. One example they use to illustrate what they have in mind here is the methodology employed by Alvarez and Asaro (1990), in which researchers collected samples of iridium to test the hypothesis that dinosaur extinction was the result of a meteorite smashing into the earth's surface (the observable implications of which would be traces of iridium in predicted layers of the earth's crust). The more general idea is, that lacking observable evidence that would represent the phenomenon in question, the careful construction of observable implications may allow researchers to associate evidence that is available with the specific assumptions of the hypothesis (in this case, that dinosaurs were killed off as a result of a giant meteorite).

Underlying their approach is the assumption that the methodology being described may provide a unified method that will lead to stronger inferences (i.e. generalizations) about the phenomena in question. For King *et al.* (1994), the ability to infer causal relations and effects is centrally dependent upon the validity and reliability of the research methodology. Validity and reliability can be enhanced, they argue:

- by articulating theories and hypotheses in terms that make specific the conditions under which the predictions of the theory may be proven right or wrong (King *et al.* 1994: 21–22);
- by recording and reporting clearly 'the process by which the data are generated' (King *et al.* 1994: 23);
- by constructing and collecting data on as many observable implications as possible (King *et al.* 1994: 24); and
- by using all relevant information in the data to generate strong inferences about causal relations (King *et al.* 1994: 26).

Like Popper (1957 [1997], 1962), the authors clearly favour an epistemology that builds upon knowledge by testing and falsifying competing claims about reality. Like Popper, they also take a critical stand on the use of history in social science research. Central to their treatment of historical and interpretive data is the idea that the original research questions and methodologies guide the selection of empirical data. For King *et al.* (1994), the challenge of incorporating history into social analysis is as follows:

> How can we make descriptive inferences about 'history as it really was' without getting lost in a sea of irrelevant detail?
>
> (King *et al.* 1994: 53)

To make sense of the past, it is incumbent upon the researcher to 'focus on the outcomes that we wish to describe or explain' (i.e. to select history according to dependent variables) and to 'simplify the information at our disposal' (King *et al.* 1994: 54). Quoting Eckstein, they suggest that 'a "case" can be defined technically as a phenomenon for which we report and interpret only a single measure on any pertinent variable' (King *et al.* 1994: 52).[93] Framed in this way, historical and other forms of qualitative data offer observations and observable implications that can be used to generate stronger inferences.

Although they concede that an appreciation of context and historical detail may help to refine the questions and concepts of social inquiry, it is clear that the questions and concepts being pursued in this context are defined and deemed relevant largely in relation to the observable implications developed through deductive reasoning. For instance, at an early point in the text the authors argue that social science research should pursue questions which are '"important" in the real world,' and 'make a specific contribution to an identifiable scholarly literature by increasing our collective ability to construct verified scientific explanations of some aspect of the world' (King *et al.* 1994: 15).[94] Later on, they warn against the 'danger' of introducing new questions or revising old hypotheses after the collection of data has begun, suggesting that questions can be posed and hypotheses formed only 'in advance' of the research process.

Similarly, Bates *et al.* (1998) argue that the methods they espouse are primarily 'problem driven,' and are not centrally concerned with building theory. However, these assertions do not appear entirely consistent with their concluding chapter (Bates *et al.* 1998), in which they assess the extent to which game theoretic models and historical narratives can be applied in other settings. Although they take great care to stress that such comparisons depend on the complementarities of the cases being compared, the very fact that they seek to assess the generalizability of their assertions suggests that they are interested in the construction of theory.

Along similar lines, Laitin (2003) argues that formal modeling and statistical analysis can (*and must*) be used to complement the contextual tapestry of historical narrative. In a particularly revealing passage Laitin (2003: 171–5) challenges Stanley Tambiah's historical account (cited in Laitin 2003) of ethnic violence among the majority Sinhalese and Tamils in Sri Lanka by demonstrating that language grievances do not correlate comparatively or historically with the onset of violence, as Tambiah's account would lead us to believe. Developing a model of language grievance (based on the Sri Lankan case), Laitin (2003) uses statistical analysis to explain why anti-Tamil legislation was subverted by Sinhalese government officials (because they had an incentive to maintain English as a *lingua franca*) and why it was the Sinhalese (i.e. those without language grievances) who first engaged in violent attacks on the Tamils. In this way, the use of statistics and formal modelling help to construct 'a new and more coherent narrative' (Laitin 2003: 174).

Much like Popper ([1957] 1997), all of these authors advocate a hypothetic-deductive approach in which the assumptions, propositions and conclusions of formal models are tested both in terms of their logical coherence and their consistency with empirically knowable facts.

An 'anti-history machine'?

The arguments being advanced by these authors mark an important departure from the assumptions offered by positivism and, indeed, by those of the neo-classical frame. However, the historiography that underlies the convenient marriage between theory and history is very different, I think, from that which informs the afore-mentioned calls for history and context. Moreover, and this is quite important, it is not at all clear that historicist interpretations and the construction of positivist theory can co-exist quite so easily. As Immergut (1998) has argued, an important tension exists between historical accounts, which stress the importance of par-ticularism, context and contingency, and the construction of theory based on the systematic comparison of historical conditions and events. Historical narratives in which social outcomes are explained in terms of a contextually specific series of conditions and events do not lend themselves well to (scientific) comparison. As she concludes:

> Without a sufficiently broad comparative perspective, historical institutionalists risk overstating the uniqueness of their case. Furthermore, it is difficult to see how such historical narratives can ever be proved wrong.
>
> (Immergut 1998: 27)

Although such assertions do not necessarily contradict the approaches favoured by Laitin (2003), Bates *et al.* (1998) and King *et al.* (1994), they do raise questions about the means by which historians and other social scientists select and interpret the 'facts' that are most important to their model. Of particular concern here is the role that history can play in social scientific research and, by extension, the way in which history is understood and applied to the understanding of social phenomena. On this question, Collingwood was deeply skeptical:

> The past, consisting of particular events in space and time which are no longer happening, cannot be apprehended by mathematical thinking ... Nor

by scientific thinking, because the truths which science discovers are known to be true by being found through observation and experiment exemplified in what we actually perceive, whereas the past has vanished and our ideas about it can never be verified as we verify our scientific hypotheses.

(Collingwood 1946 [1992: 5])

To Collingwood (1946 [1992]) and to historians of his persuasion (e.g. Carr 1951; Hobsbawm 1987 [1989]; Moore 1966), critical reflection about the act of historical interpretation constitutes a central aspect of post-scientific revolution thinking about the past. In this respect, a critical reflection about the core values and assumptions that underlie social science research and social science questions is fundamental to the acquisition of knowledge. What makes the propositions being advanced by Laitin (2003), Bates *et al.* (1998) and King *et al.* (1994) particularly problematic is the relative lack of theorizing about the core values on which history and social science *should be* based.[95]

A related concern is that power and influence – as opposed to open and critical reflection and debate – will define and decide the direction of scholarship. Towards the end of their concluding chapter, Campbell and Pedersen (2001b) make the crucial observation that 'political criteria' (alongside empirical validity and normative considerations) will determine the extent to which different theoretical 'paradigms' in the institutional literature will be able to combine to produce a second movement in institutional analysis:

There is ample evidence showing that those paradigmatic views that came to dominate the intellectual landscape at different moments in history did so in part because they were backed by substantial material resources and intellectual elites who were able to gain footholds in important institutional arenas where they could articulate their ideas, train protégés, and establish influential intellectual and professional networks for the propagation of their views.

(Campbell and Pedersen, 2001b: 247)

In his critique of Bent Flyvbjerg's (2001) call for a more contextualized social science, David Laitin argues that methodologies based on context and historical narrative 'must be combined with statistical and formal analysis if the goal is valid social knowledge' (Laitin 2003: 170). Responding to the 'Perestroikan challenge' to the perceived lack of pluralism within the American Political Science Association, Laitin concludes that:

A scientific frame would lead us to expect that certain fields will become defunct, certain debates dead, and certain methods antiquated. A pluralism that shelters defunct practitioners cannot be scientifically justified.

(Laitin 2003: 180)

In other words, the explanatory utility of a methodology or a discipline can be measured in terms of the contribution it makes to our understanding of social phenomena. Scientific rigor thus provides a 'natural' means of selecting the methodologies that will best explain the things we want to know. However, what we want to know and how we go about knowing it are surely contestable questions, whose values and assumptions should be subject to the kinds of hard inquiry that Popper (1957 [1997]) so strongly favoured. Without a strong and contested justification of what constitutes desirable social knowledge, the difference between using scientific rigor to separate good theories from bad and the construction of dogma becomes very fine indeed.

In short, ambitious attempts to merge scientific approaches with historical narratives are limited in the sense that they subvert the peculiarities of historical events to the logic of deductive reasoning. Without an explicit justification for the selection and interpretation of historical 'facts' (and for what constitutes facts), treatments of this kind are unsurprisingly threatening to those who understand history (including the questions, assumptions, values and methodologies that inform one's history) as contingent and context-specific.

A second way of incorporating history into a comparative frame is one that rejects the covering laws of hypothetic-deductive analysis and employs instead an inductive analysis of historically contingent processes and events.

Bringing history back in: advancing knowledge for social change

Calls for comparison in social science often invoke the idea that Weber may provide a more viable means of combining the rigor of social theory with the context and nuance of historical social analysis (Moore 1966; Skocpol 1979; Evans 1995; Kohli 2004). However, as Vandergeest and Buttel (1988) have argued, Weber offers a number of different interpretations about the nature of social processes, and may therefore be interpreted in a number of different ways. One 'version,' they suggest, 'consists of fitting particular cases into a pre-given model of society' (Vandergeest and Buttel 1988: 684). A second version – exemplified in the work of Karl Polanyi (1957), E. P. Thompson (1963) and Barrington Moore (1966) – suggests that 'grounded' historical analysis may provide the basis for comparative social analysis.

Underlying this second Weberian tradition are a number of assumptions about the nature of reality and the construction of knowledge. One assumption is that social structures may be used to theorize and compare different social processes and outcomes. As Philippa Bevan (2004) has argued, the assumption here is that 'durable and pervasive structures' exist, and they 'can be theorised across a range of instances' (Bevan 2004: 8). A classic case in point would be marriage, an institution which is practiced in many societies, but one whose nature varies enormously on the basis of formal rules, laws and policies and 'informal' traditions and customs.

A second assumption is that language and concepts may provide a (relatively) accurate view of the world around us. Framed in this way, reality 'has a consistently identifiable nature, and hence is imbued with inherent causal powers that can be represented indirectly by concepts' (Morrow and Brown 1994: 137).

A third assumption is that the questions and problems of social analysis are empirically and historically driven. In *The Protestant Ethic and the Spirit of Capitalism*, for instance, Weber (1958) starts from the empirical observation that the vast majority of wealth in the North American and Western European economies (at the end of the nineteenth century) tended to be concentrated in the hands of Protestant families and communities. The central aim of Weber's study is therefore to understand the social and religious factors that explain this seemingly widespread phenomenon.

A fourth and crucial assumption is that the theories, hypotheses and assumptions of social analysis are open to questions of history, contingency and context. Framed in this way, 'theory is not used to "predict," but is employed in dialogue with evidence and observation to construct an analytic account and analysis of what is and what might be possible' (Vandergeest and Buttel 1988: 688).

The basic idea is therefore that research entails a dynamic process whereby initial theories or 'conjectures' are refined and revised as the research unfolds. It also aims to infer and to build theory on the basis of structured comparative analysis.

What one compares, and how one decides the ontological boundaries of what one compares (i.e. what is a country, a society, etc.) is of course an open question. In political science, for instance, the nation-state (which includes, *inter alia*, the executive, the legislature, the bureaucracy, the military, the police and the judiciary) is commonly used to provide the primary unit of analysis. An alternative method rejects the ontological importance of national boundaries, and embraces instead the idea that individual cases (e.g. countries, societies, etc.) may be understood in relation to a wider system or whole. For instance, Philip McMichael (2000) suggests a method of 'incorporated comparison' in which the ontological boundaries of national states and histories are framed in relation to a wider historical system and process. Similarly, Charles Tilly (1984) has theorized a kind of 'encompassing comparison' in which all instances may be understood and explained in relation to a single, general system (cf. Wallerstein 1974).

How one compares is also a point of potential difference and debate. Guy Peters (1998), for instance, identifies six different ways in which scholars (primarily political scientists) may engage in comparative analysis:

- single country studies (by which he means single case studies that are compared with other cases);
- process studies (e.g. revolutions);
- institution studies (e.g. legislatures);
- 'typographical studies,' (essentially Weberian ideal types);
- regional studies (two or more countries);
- global statistical studies ('large N').

Part of the challenge of doing comparison is that the field is defined with such breadth and ambiguity that it can include a wide variety of perspectives and approaches. (Many, for instance, would disagree with Peters that case studies can or should be included on the list). There is also the problem that aspects of what we might compare (e.g. judicial systems, education policies, central banks, etc.), which are assumed to be different or unique may be 'polluted' or affected by external/ transnational factors and processes, suggesting an ontological blurring of what may have been perceived as purely 'domestic' and 'international' factors and processes. For comparative scholars, a challenge of 'doing comparison' therefore entails the problem of accounting for the multiple ways in which transnational forces and institutions may influence domestic politics, and vice versa.

In his classic book, *Big Structures, Large Processes, Huge Comparisons*, Charles Tilly (1984) argues that comparisons may be distinguished among four 'types' (ideal types, as it were):

- *individualizing comparison*, in which each case is essentially unique;
- *variation-finding comparison*, in which 'many forms of the phenomenon exist,';
- *encompassing comparison*, in which all instances may be understood and explained in relation to a single, general system (cf. Wallerstein 1974); and
- *universalizing comparison*, in which 'common properties' exist 'among all instances of a phenomenon.'

Tilly's basic point is that the typology does not depend on the 'strict internal logic' of comparison (as positivism would suggest). Rather, it offers a heuristic that helps to make sense of how and why large comparative inferences may be made.[96]

As noted earlier, the preceding chapters describe a field that has become highly dependent on single country and case study research. Using Tilly's typology, many of these would conform to what he is calling 'individualizing comparison,' in which every case is taken to be unique or 'encompassing comparison' in which single or local case studies are framed in relation to broader systemic processes and events, such as globalization, structural adjustment or the green revolution. Far less developed, I would argue, is what Tilly is calling universalizing and variation-finding forms of comparison.

To illustrate what I have in mind here, let us consider briefly three examples of the ways in which history may frame and inform the comparison of 'big structures and large processes.' One is *Embedded Autonomy*, by Peter Evans (1995).

A central aim in Evans' three-country study is to establish the conditions under which peripheral economies may undergo processes of industrial transformation and economic growth. Drawing upon developmental state theory, a central focus is on questions of state capacity and state-society relations in the developing world. His principal assertion is that effective state intervention required a 'concrete set of social ties that binds the state to society and provides institutionalized channels for the continual negotiation and renegotiation of goals and policies' (Evans 1995: 12).

Following Weber, his methodology is primarily geared towards an inductive search which 'starts with contextual differences and then looks for underlying regularities' (Evans 1995: 29). For Evans, the central challenge is to organize (and to simplify) the analysis of what are in fact three very different countries: India, South Korea and Brazil. The strategy, he suggests, is

> ... to start by constructing two historically grounded ideal types: predatory and developmental states ... Predatory states extract at the expense of society, undercutting development ... Developmental states not only have presided over industrial transformation, but can be plausibly argued to have played a role in making it happen.
>
> (Evans 1995: 12)

The first point to make about Evans' research strategy is that it is rooted in a specific history of state formation, which offers a heuristic for understanding other cases in his study. His rendering of the predatory state, for instance, is based primarily on the political history of Zaire (Evans 1995: Chapter 3). Likewise, his understanding of developmental state properties is derived primarily from the historical literature on Japan (Johnson 1982). The second point is that his understanding of ideal types allows for variation that informs and expands his terms of analysis. Alongside his two 'ideal types,' for instance, he further theorizes the nature of state-society relations and regime type on the basis of Korea and Taiwan. A final point is that his analysis cuts across a wide and ambitious empirical terrain without lapsing into 'glorified empiricism.' 'To fulfill the potential of a comparative institutional approach,' he argues,

> The Weberian hypothesis must be explored across agencies and countries ... The key is to identify differences in the way states are organized and then connect these differences to variations in development outcomes.
>
> (Evans, 1995: 40–1)

In short, Evans (1995) provides a systematic way of using history and context to theorize and establish variations concerning the ways in which states may foster processes of industrialization and development.

Similarly and building upon Evans (1995), Atul Kohli (2004) uses three 'ideal types' of state capacity to understand processes of industrialization and economic growth in India, Nigeria, South Korea and Brazil:

- *Neopatrimonial states*, having 'weakly centralized and barely legitimate authority structures, personalistic leaders unconstrained by norms or institutions, and bureaucracies of poor quality' (Kohli 2004: 9);
- *Cohesive-capitalist states*, in which effective state bureaucracies have established strong and productive relations with business and labour;
- *Fragmented-multiclass states*, in which public authority and political power is dependent upon a wider (and therefore) more fragmented coalition of social and class interests.

Again, like Evans, Kohli's conceptualization is rooted in the idea that the individual histories of India, Nigeria, South Korea and Brazil may provide the empirical basis for comparison and generalization. The value of this approach is its parsimony and its attention to historical detail. Unlike the 'combined approaches' described earlier, it rejects the idea that social relations may be theorized on the basis of a hypothetic-deductive model and draws its theories instead from particular, long-term processes and events.

A final and more recent contribution is *Social Democracy in the Global Periphery* by Richard Sandbrook, Marc Edelman, Patrick Heller and Judith Teichman (2007). Drawing directly upon Kohli (2004), Evans (1995) and other state theorists, Sandbrook *et al.* (2007) aim to understand the historical conditions under which social democracies in the developing world have been able to reconcile 'the exigencies of achieving growth through globalized markets with extensions of political, social and economic rights' (Sandbrook *et al.* 2007: 3). Their central claim, supported by rich historical analysis, is that social democratic rights are most extensive in the instances during which lower-class mobilizations have led to broader political coalitions, which push governments into processes of social democratic reform.

Like Evans and Kohli, their frame is broad and ambitious, involving a comparative analysis of four peripheral states: Costa Rica, Mauritius, Chile and the Indian state of Kerala. It also employs a number of innovations not commonly found in political science, or in development. First, they abandon the assumption (common in comparative politics) that the only or primary unit of analysis can or should be the nation-state. Instead, they offer a comparison of three nation-states (Costa Rica, Mauritius and Chile) and one sub-national state (i.e. the Indian state of Kerala), suggesting that the crucial point of analysis and comparison concerns the constellation of power and authority governing the determination and implementation of social and economic policy.[97]

Second, their selection of states is one that moves away from the traditional focus on industrialization and economic growth (e.g. Korea, India, Brazil) to ones whose record of achievement rests in the ability to improve human development, measured in terms of primary healthcare, universal education, social security, poverty reduction and democratic reform.

Third, their methodology provides an unusually creative and collaborative way of approaching interdisciplinary research. According to their Acknowledgments (Sandbrook *et al.* 2007: vii), the book 'emerged from a movable seminar,' which involved a series of five public symposia. Although they are from different disciplinary backgrounds (two political scientists, one sociologist and one anthropologist), the authors were able to assemble a clear and coherent account that achieves strong inter-disciplinary insights about the conditions under which states and social movements may extend social, political and economic rights.

Fourth and crucially, their treatment of history employs a fascinating variation on the Weberian theme. Instead of classifying agencies, states or coalitions in terms of ideal types, the authors suggest that *historical causal factors* may be divided into three separate categories:

- *Structural factors*, shared by all four cases, involving 'early and deep, albeit dependent, integration into the global capitalist economy' (Sandbrook *et al.* 2007: 30);

- *Configurational factors*, concerning the nature and alignment of class forces, 'the most propitious pattern being one that weakens the landlords while strengthening the working and middle classes' (Sandbrook *et al.* 2007: 31); and
- *Conjunctural factors*, 'critical junctures in a country's history, in which social actors, through political struggles, propel societies down a particular path' Sandbrook *et al.* 2007: 31).

In so doing, they provide an effective means of structuring and therefore comparing their cases across space and time. But they also open and leave room for the role of politics in social analysis.

These are but three examples of studies that capture the nuance and details of history while at the same time advancing a scholarship that moves beyond the perceived limitations of positivist analysis and case study research. There are, of course, many other ways in which comparative research may be theorized and pursued in the context of development. David Landes (1998), for instance, offers a particularly grand effort to theorize and explain through history the 'wealth and poverty of nations.' Similarly, John Harriss (2000) provides an innovative attempt to theorize and compare political and economic outcomes across the Indian states.

The preceding analysis suggests that studies of this kind have been few and far between (or at the very least, they have not featured prominently in the leading development journals). For scholars entering the field (and for older ones, too) the opportunities appear to be vast.

Concluding remarks

To conclude the book, I now want to make a few final points about the normative, epistemic and historical implications raised (but not necessarily resolved) in this analysis.

One point concerns the relationship between social theory and social action. As the preceding chapters suggest, the idea of connecting social theory with social action has lost considerable ideological appeal. At a time when serious doubts were being raised about the ability of states and social movements to engineer universally desirable forms of progress, it is perhaps no surprise that social science researchers have become increasingly wedded to theories and worldviews that reject the grand sweep of history in favour of ones that collected and ascribed new meaning to the aggregation of individual needs and preferences. In *Making Social Science Matter*, for instance, Bent Flyvbjerg (2001) argues that the social sciences have moved away from what he calls 'value-rationality,' in which the ideals of natural science have usurped a more traditional Aristotelian concern for questions concerning 'Where are we going?'; 'Is this desirable?' and 'What should be done?' (Flyvbjerg 2001: 60).

The reasons for the normative retreat in American political science are a matter of some debate and go well beyond the scope of this book. Included among many possible explanations are the professionalization, fragmentation and organization of academia (Cohn 1999; Ricci 1984); the development of new and powerful quantitative techniques (Cohn 1999); the desire to emulate natural science, particularly among mainstream economics (Fine 2001; Flyvbjerg 2001; Cohn 1999); the aversion to value-laden theorizing in America during McCarthyism and the Cold War (Leys 1996); the pragmatic orientation of development studies (Leys 1996); and, more recently, the widespread discrediting of grand normative theories of development and change (Gore 2000; Leys 1996; Schuurman 1993).

Within neo-classical economics, the power to make generalizations lies in its commitment to quantification (i.e. assigning numerical values to stated preferences) and statistical analysis. Outside of economics, the perceived advantage of using neo-classical theory to understand social phenomena rests in its ability to develop, on the basis of hypothetic-deductive models of individual decision-making, theoretical propositions about the conditions under which individuals, groups and societies will provide for themselves collective goods, such as irrigation, literacy, markets, democracy and good government.

As Ben Fine (2001) has argued, the ability of social capital to appeal to such large numbers of agencies and interests rests in its ambiguity and therefore its ability to occupy a 'middle ground' (what Fine calls a 'scholarly third wayism') that gives the impression of being receptive to questions of history and difference while at the same time offering an analysis that goes beyond 'mere description' (Fine 2001: 190). Framed in this way, a large part of its appeal stems from a desire on the part of economists to incorporate 'non-economic' factors into their analysis and the desire of non-economists to adopt the language of methodological individualism and rational choice.

But the attraction also reflects a desire on the part of development and academia in general to invent and pursue new and increasingly fashionable trends and concepts, illustrating what he suggests is 'a more general trend towards the popularisation and degradation of scholarship' (Fine 2001: 191). This degradation, he concludes, reflects the inability of development scholars (and of social scientists in general) to reconcile the 'postmodern' critique that all knowledge is biased and partial and a positivist desire to understand and describe reality:

> Where postmodernism has departed the material and objective for the symbolic and the subjective, so its alter ego in more traditional social science, rapidly being subsumed under social capital, has hardened in its use of universal analytic categories in order to address what is presumed to be an unproblematic descriptive and statistical reality.
>
> (Fine 2001: 193)

Whether postmodernism facilitated the neo-classical turn, we can certainly detect what appears to be an uncomfortable consistency between the fragmentation of politics, knowledge and reality, and the fragmentation, dislocation and globalization of economic production and social life that begins to take place during the 1960s and 1970s (Harvey, 1990). Frederic Jameson (1984) makes this link explicit when he argues that:

> ... aesthetic production today has become integrated into commodity production generally: the frantic economic urgency of producing fresh waves of ever more novel-seeming goods (from clothing to airplanes), at ever greater rates of turnover, now assigns an increasingly essential structural function and position to aesthetic innovation and experimentation.
>
> (Jameson 1984: 56)

Similarly, David Harvey (1990) suggests that postmodernism represents a break with high modernist traditions in literature, architecture, scholarship and art, stemming primarily from the technological 'compression' of time and place in the context of capitalist social relations:

> Aesthetic and cultural practices are peculiarly susceptible to the changing experience of space and time precisely because they entail the construction of spatial representations and artefacts out of the flow of human experience ... The experience of time and space has changed, the confidence in the association between scientific and moral judgments has collapsed, aesthetics has triumphed over ethics as a prime focus of social and intellectual concern, images dominate narratives, ephemerality and fragmentation take precedence over eternal truths and unified politics, and explanations have shifted from the realm of material and political-economic groundings towards a consideration of autonomous cultural and political practices.
>
> (Harvey 1990: 327–8)

Framed in this way, the fragmentation of theory and reality reflects a wider fragmentation of economic and social life (cf. Corbridge 1990).

A second point of conclusion concerns *the viability* of connecting theory and praxis. For scholars wedded to the idea of using class analysis to advance a more 'radical' social agenda, my emphasis on methodology, comparison *and Weber* will no doubt fall on deaf ears. Indeed, some forms of Weberian analysis have been criticized for aiming simply to understand the world, 'without an explicit agenda or politics' (Vandergeest and Buttel 1988: 690; cf. Buttel and McMichael 1994). To this critique, I would suggest that there is nothing in the methodology being advanced that would preclude the use of class analysis for social ends. Indeed, the analysis offered by Sandbrook *et al.* (2007) illustrates the powerful ways in which Weberian theory and class analysis may be combined to challenge conventional wisdoms about comparison, social democracy and development.

A final point concerns the challenge and danger of turning theories into action. As noted in Chapter One, development is a field that engenders *very* strong feelings about poverty, suffering, inequality and injustice, creating strong expectations that the study of development be intimately and essentially connected with the practice of development. The notion that theory may be assessed only or primarily in terms of its ability to provide answers and solutions to the world it 'exposes' is of course a child of the Enlightenment, and one that significantly underplays the challenge and danger of turning particular theories and ideas into action. Indeed, if Foucault teaches us anything, it is that we be wary of theories that turn people into projects, and freedom into 'theoretical formulas' defined primarily or entirely in relation to the needs of a system or plan (Flyvbjerg 2001). Therefore, the challenge is to develop a perspective (or, dare we say, a paradigm) that can incorporate and protect the needs and perspectives of 'distant strangers' without lapsing into a relativism that denies the ability of scholars, activists, etc. to represent or compare other cultures, societies and people.

As Thomas Kuhn (1962) reminds us, paradigms consist of intellectual and institutional elements, which represent the axiomatic principles (the core questions, values, assumptions and methodologies) on which 'normal science' is based, and the organizational structures, incentives and practices in which its practitioners are engaged. Framed in this way, facts and ways of knowing are made sensible in relation to broader cultures or 'paradigms' of knowledge and science. According to Kuhn (1962), scientific revolutions occur when the application of existing theories and methodologies produces consistently unexpected results, which existing theories are unable – or unwilling – to explain. During such periods, he argues, radically new ways of posing and understanding these questions begin to emerge. However, the organizations and incentive structures that underlie modern science do not always or necessarily lend themselves to open and scholarly debate. On the contrary, the preservation of disciplines, methods and careers within and among these disciplines may produce very intense struggles over the construction of knowledge.

The evidence and argument presented in this book provide grounds for pessimism and optimism about the possibilities of transcending the current state of affairs. On one hand, the theories, concepts and methodologies of neo-classical theory appear to occupy the commanding heights of development theory and practice (Kanbur 2002). On the other, the evidence considered in this book (especially concerning the history of ideas) appears to suggest that the events and contradictions of history and human experience (e.g. the 'lost revolution,' the Holocaust, 1968) may combine to unravel theories and worldviews that no longer conform to the moral, intellectual and aesthetic norms of the time. As James Scott (1998) has argued, the 'availability of knowledge ... depends greatly on the social structure of the society and the advantages that a monopoly in some forms of knowledge can confer' (Scott 1998: 334).

Notes

1 For a good indication of how far this rhetoric can go, readers may want to consult *The Lords of Poverty* (1989) by Graham Hancock. More firmly rooted in neo-classical thinking about markets and the limitations of central planning, *The White Man's Burden* (2006) by William Easterly offers a similar chorus of condemnation. More lucid in their understanding and explanation of development failures are *The Anti-Politics Machine* (1994) by James Ferguson and *Cultivating Development* (2005) by David Mosse.

2 Peet and Hartwick (1999: Chapter 2), Fine (2001), Gilpin (2001, Chapter 3), Evans (1995: Chapter 2), Brohmann (1995), Kanbur (2002), Bracking (2004) and Sumner (2007).

3 Hilary Putnam (2002: 49, citing Sen) has argued that rationality in 'mainline' economic theory implies either 'internal consistency of choice' or 'maximization of self-interest.'

4 As Kanbur (2002) has argued, the assumption in 'mainstream economics' is that parties to an economic transaction will find mutual benefit if and when they are acting voluntarily and on the basis of accurate information, and if and when they have 'sufficient alternative parties' (Kanbur 2002: 478) with whom they may transact. An important concept here is 'marginal utility': the idea that the value of a good or service is not a reflection of the amount of labour that goes into its production (as both Marx and Adam Smith would lead us to believe), but of preferences and consumption decisions made in relation to the real and perceived supply and therefore scarcity of goods and services (Peet and Hartwick 1999: Chapter 2).

5 See especially, Harriss *et al.* (1995), Bates (1995) and Hodgson (1993). For treatments in political science, see Immergut (1998), Hall and Taylor (1996) and, more recently, the collection of essays in Campbell and Pedersen (2001).

6 Here it is worth adding that the assertions being made by Spivak (1988) and others (e.g. Kapoor 2004; Goss 1996) were directed principally towards universities and intellectuals engaging in research provided for and by the academy. However, if we go beyond the academy, such assertions are equally or even more difficult for 'the industry' of NGOs, think-tanks and consultancies engaged in development research.

7 I am extremely grateful to Alex Parisien for collecting, coding and entering the data for this survey.

8 I am also aware that comparing countries provides only one means of comparison and has its limitations, which we explore in Chapter Six.

9 One important form of power and legitimacy came from peasant rebellions in Asia (e.g. China and Indochina) and Africa (e.g. Kenya), whose emphasis on landlessness, rural tenancy and agrarian reform was (for a time) broadly consistent with communist goals of social re-distribution. Whether peasants in Africa and Asia were demanding the collectivization that would follow the establishment of socialist republics in Russia, China and Vietnam, the key point to emphasize here is that the peasant rebellions of the twentieth century had a strong re-distributive agenda and that communism (for a time) helped to capture and reflect these demands.

10 They were 'new' in the sense that they aimed to move beyond the universalizing assumptions and analytic categories of class and class struggle, emphasizing instead the historically and culturally contingent ways in which social actors perceive and engage in different forms of social mobilization.

11 Although post-structuralism and postmodernism 'are not identical,' writes Pauline Marie Rosenau (1992: 1), 'few efforts have been made to distinguish the two, probably because the differences appear to be of little consequence.' Similarly, Sarup (1989: 131) suggests that 'There are so many similarities between post-structuralist theories and postmodernist practices that it is difficult to make a clear distinction between them.'

12 See, for instance, Harvey (1990), Jameson (1997), Brass (1991), Sarup (1988), Eagleton (1997), Bellamy Foster (1997) and Meiksins Wood (1997). For critical summaries see Carter (1997) and Rosenau (1992: Chapter 8).

13 To be clear, I am not making the case that historical approaches do not exist in the commons literature. Indeed, a search on Indiana University's comprehensive CPR search engine (http://www.indiana.edu/~iascp/Iforms/searchcpr.html) produced more than 50 titles relating to 'history and the commons' (accessed 7 May 2003). Classic historical treatments of the commons would include Netting (1981), Dahlman (1980) and Acheson (1988).

14 Although he shared with positivism a disdain for the untestable, Popper was, in a number of ways, a 'realist' or a 'rationalist' in the sense that he gave value to the existence of unobservable theoretical entities. As Hacking (1983: 43) has argued, he believes that 'untestable metaphysical speculation is the first stage in the formulation of more testable bold conjectures.' Popper is sometimes associated with positivism, Hacking (1983: 43) argues, because he 'believes in the unity of the scientific method.'

15 A related assertion is that the selection of cases fails to conform to standardized norms about bias and sampling. George and Bennett, for instance, suggest that selection biases 'can occur when cases or subjects are self-selected or when the researcher unwittingly selects cases that represent a truncated sample along the dependent variable of the relevant population of cases.' Selecting cases on the basis of 'the dependent variable' (i.e. the outcome the research is aiming to explain), they argue, 'can help identify which variables are not necessary or sufficient conditions for the selected outcome' (George and Bennett 2004: 23). However, researchers must be careful that the selection is not biased in favour of cases that would either confirm or falsify the original hypothesis. A final and related concern is that case study research generates data that are insufficiently representative to inform a wider class of phenomena.

16 Along similar lines, Alexander George and Andrew Bennett (2004: 12–14) have argued that the methodologies advanced by King *et al.* (1994) understate: (1) the challenge of capturing complex causal relations; (2) the ability of social science research to generate new questions and hypotheses; and (3) 'the dangers of "conceptual stretching" that can arise if the means of increasing observations include applying theories to new cases, changing the measures of variables, or both,' (George and Bennett 2004: 13).

17 Although non-renewable natural resources are potentially depletable, they are ultimately resilient in the sense that the costs of finding and extracting them are proportional to their perceived availability (Homer-Dixon 1991; Simon 1981). As Simon (1981) argues, individuals will never pump 'the last barrel of oil' because the costs of finding and pumping it will always far exceed the marginal benefits of actually using it. Renewable resources differ in the sense that they *usually* regenerate themselves, but the rate at which they undergo this process can vary.

18 In so doing, commons scholars (such as Ostrom 1990; and the collection of essays in McCay and Acheson 1987; and Bromley *et al.* 1992) provided an important critique to Hardin's assertion that only state intervention or private property would resolve the tragedy of the commons, which at the time had become extremely influential. I am grateful to an anonymous reviewer for highlighting this point.

19   Although Hardin's article (1968) did little to dispel Malthusian biases, it did invoke a new body of scholarship that called into question the notion that the commons will always be open access, thereby dismantling the deterministic idea that individuals will always degrade the commons.

20   Seminal works in this literature would include Ostrom (1990), Wade (1988), Baland and Platteau (1996), Uphoff *et al.* (1990) and Bromley (1992), as well as the collection of essays in McCay and Acheson (1987), Bromley *et al.* (1992) and, more recently, Dietz *et al.* (2002).

21   Mosse (1997) classifies these two schools in terms of 'economic-institutional' and 'sociological-historical' explanations. Goldman (1998) refers to 'tragedy' and 'anti-tragedy' scholars, and separates the latter into 'human ecologists,' 'development experts' and 'global resource managers.'

22   Similarly, Douglass North contends that changes in the costs of obtaining and synthesizing information, factor (land, labour and capital) prices and technology are the 'most fundamental source' of institutional change (1990: 83–4).

23   These issues are addressed in some detail by Blair (1996), who considers the extent to which Hardin's three 'solutions' to the commons dilemma were consistent with broad principles of democracy. Drawing upon the case of local forestry in India, he argues that democracy is 'quite compatible' (Blair 1996: 493) with privatization and centralization, but somewhat less so with local control. User groups which can restrict membership to those who follow the rules 'can be democratic, in the sense of participation and accountability,' (Blair 1996: 494). Blair says little about what democracy would entail in this instance, but in an earlier footnote defines it as:

> a governance system comprising three elements: *popular sovereignty* in that the state is both accountable and accessible to its citizens; political equality, in that citizens enjoy full human as well as full legal rights; and political liberty in the form of freedom of speech and assembly …
>
> (Blair 1996: 477; emphasis in original).

Precisely how Leviathan would be 'accommodated within a democratic framework' Blair (1996: 482) does not specify, beyond the fairly broad generalization that there were some theoretical consistencies (about the Hobbesian state of nature, and the need for an authoritarian state) between Hardin and Hobbes. More germane to our purposes, there is little (explicit) treatment of the difficult relationship that is known to exist between democracy and social inequality (an issue explored at great lengths by democratic theorists like Macpherson 1978).

24   Contradicting this position is the notion that individuals are less likely to act collectively when the distribution of endowments and entitlements is relatively skewed (Ostrom 1990: 119, 211; Ostrom *et al.* 1993: 86; Putnam 1993: 174–6; Libecap 1995: 168; Mitchell 1995: 247). Putnam (1993) takes this one step further and argues that social inequality actually undermines the prospects for successful collective action. This he does by making a conceptual distinction between 'vertical' hierarchies, based on reciprocal patron-client ties, and 'horizontal' organizations in which individuals interact on relatively equal footing (Putnam 1993: 89–97, 173–5). Two propositions support his case. First, he argues, hierarchies stifle the production of reliable information, which, in turn, undermines the distribution of social reputation (Putnam, 1993: 173). Second, power inequalities reinforce norms of superiority and subordination, perpetuating a relationship based on dependency and control (Putnam 1993: 107–11). Under these conditions, he concludes, exploitation and mistrust become the norm, reinforcing the expectation of opportunism. Whether and how structures affect individual incentives are of course highly dependent on the historical forces that give rise to these structures in the first place. As James Scott (1976: 202) has argued, for instance, collective action to counter the effects of market capitalism happens in differentiated villages not necessarily or only because they are differentiated, but because differentiated villages tend to be closer to market centres, 'where commercial forces have been strongest.'

25 The categorization for this table is drawn from the very useful review (of a different literature) in Campbell and Pedersen (2001).

26 Wade's explicit treatment of poverty and inequality may lead some to conclude that he has more in common with what I am calling the entitlement school than he does with theorists of collective action. In response to this, I would argue that although he does explore the extent to which common property institutions (namely common irrigation and field guarding) provide tangible benefits to the rural poor (see especially, Wade 1988: Chapter 6), he is principally concerned with the conditions under which 'some peasant villagers in one part of India act collectively to provide goods and services which they all need and cannot provide for themselves individually' (Wade 1988: 1). Moreover, and as this quotation suggests, his explanatory framework is firmly grounded in a tradition of methodological individualism and rational choice, which is very consistent with that employed by other collective action scholars, such as Ostrom (1990) and Baland and Platteau (1996).

27 Indeed, the positions taken within this debate have much in common with those of the 'growth versus inequality' debates, which questioned the trickle-down effect of a 'rising tide,' of the 1990s.

28 Note that Wade (1988: 112–3) reaches similar conclusions about the social composition of the village councils in 'Kottapalle.'

29 Conflict of course does play a role in the collective action literature, but it is most commonly understood in terms of a bargaining scenario, in which individuals and groups negotiate and pursue strategies that will best meet their individual and collective interests. See especially, Ostrom 1990; Ostrom *et al.* 1994.

30 Such entitlements can, of course, be centrally dependent on rights of private property, illustrating the important and overlapping relations that can exist among different forms of property rights regime (e.g. private, common, state). I am grateful to an anonymous reviewer for highlighting this important point.

31 Although collective action scholars (such as Wade 1988; Ostrom 1990 and Baland and Platteau 1996) go to great lengths to undermine the assertion that private property rights will always produce a more efficient allocation of social cost, their emphasis on institutional arrangements and individual incentives are, in fact, very consistent with Hardin's argument in favour of private property. Where they differ of course is on the means by which these incentives can or should be delivered. For the new institutionalists, individual incentives are provided not (necessarily) by rights of private property – or by the state – but by the individual benefits which can be gained (or risks avoided) by supporting common property regimes. Whether these incentives arise as a result of enclosure, encroachment or expropriation (i.e. mechanisms of change historically associated with the development of market capitalism) is largely subordinate to the question of whether they result in an efficient allocation of social cost (cf. Goldman 1998; Thompson 1998).

32 Although he consciously avoids the task of defining 'efficiency, equity and sustainability,' it is clear from his analysis that Agrawal is principally concerned with 'the durability of institutions' (Agrawal 2001: 1650) and the conditions under which individuals will implement and enforce common property regimes.

33 Since first writing this paper, I have been informed that Agrawal has moved to embrace a more historical and far less positivist approach to the study of the commons. Moreover, it is worth pointing out that his own body of scholarship (especially Agrawal 1999 and 2005) moves far beyond the kinds of positivism being prescribed in Agrawal (2001). That having been said, I would argue that, for the purposes of this exercise, the articulation matters more than the individual. In other words, even though the author may have changed his views, the publication (i.e. Agrawal 2001) still constitutes an important articulation in an influential and widely read journal.

34 From http://www.indiana.edu/~iascp/Iforms/searchcpr.html, accessed 7 May 2003.

35 Indeed, collective action theorists have criticized Hardin (1968 [2005]) for the inflexibility of his prescriptions, but – remarkably – not for the ethical positions he takes

on the subversion of individual freedom, which he defended in the name of collective survival. On the question of population, for instance, Hardin (1968 [2005: 31]) was abundantly clear:

> To couple the concept of the freedom to breed with the belief that everyone born has an equal right to the commons is to lock the world into a tragic course of action.

36 Indeed, interpretations of 'the science of Marx' became a science in its own right. In his Introduction to the *Communist Manifesto*, Gareth Stedman Jones (2002: 121) suggests that when writing *A Contribution to the Critique of Political Economy*, Marx began to present his work as 'a form of disinterested scientific inquiry,' responding in part to the repressive state of politics ongoing at the time in Prussia.

37 'As long as people can lay their hands on the means of production (tools, resources, land),' Wolf (1997: 77) argues, 'and use these to supply their own sustenance – under whatever social arrangements – there is no compelling reason for them to sell their capacity to work to someone else.'

38 Marx made an important distinction between 'use' and 'exchange-values' of commodities that satisfy human wants. 'Every useful thing,' Marx argued (in *Capital, Volume I*, cited in Freedman 1961: 27), 'is an assemblage of many properties, and may therefore be of use in various ways.' 'But this utility is not a thing of air … Use-values become a reality only by use or consumption' (*Capital, Volume I*, cited in Freedman, 1961: 28). 'Exchange-value' on the other hand constitutes the quantitative embodiment produced when 'values of one sort are exchanged for values of another sort, a relation constantly changing with time and space,' (*Capital, Volume I*, cited in Freedman 1961: 28). Irrespective of the 'use-values' they acquire, Marx argued, all commodities share one 'common property' – that they are products of human labour. 'A use-value, or useful article, therefore, has value only because human labour in the abstract has been embodied or materialized in it' (*Capital, Volume I*, cited in Freedman 1961: 30).

39 That Marx was consistent *or consistently deterministic* in his writing is of course a matter of some debate. For a useful synthesis and interpretation of this debate, see Giddens (1971: Chapter 2) and Giddens (1979: Chapter 4); and Morrow and Brown (1994). Louis Althusser (1969), for instance, argued that *The German Ideology* marked an important 'break' in which Marx moved away from the idea that alienation was the product of a timeless and linear evolution, and was instead a historical phenomenon, 'which can only be understood in terms of the development of specific social formations' (Giddens 1971: 19).

40 Actually, he theorized two: a 'Prussian path' in which capitalist differentiation occurs 'from within a landlord economy'; and an American one, in which 'there is no landlord class' (Harriss 1982: 123).

41 And as Djurfeldt (1982: 151) points out, Chayanov's thesis provided an alternative interpretation of the very same data Lenin was using to describe a process of capitalist transition; from Chayanov's perspective, the 'differentiation' Lenin perceived in the data was instead a cyclical (over the life cycle of a family) and demographic (relating to family size) variation of investment in land and labour, not a polarization of agrarian classes.

42 For List, a central area of concern was the ability of the nation state to use tariffs and trade protection to foster infant industries capable of competing with other industrialized economies (Chang 2002). Similarly, the developmental state literature highlighted the ability of governments in South Korea, Taiwan and Japan to manage the entry of their economies into the world economy by protecting infant industries through the use and gradual removal of tariffs, subsidies and other forms of infant industry protection (Rapley 2002; Kohli 2004).

43 Referring specifically to the work of Heidegger, they further differentiate between 'Division I' hermeneutics, which emphasizes 'everyday practices and discourse, which is overlooked by the practitioners, but which they would recognize if it were pointed out to them,' and 'Division II' hermeneutics, in which truth and reality are hidden – or masked – by everyday discourse and practice (Dreyfus and Rabinow 1983: xxi-xxii).

44 Underlying this position was the notion that human experience is essentially subjective and sublime, and it therefore defies any form of description or representation (Rosenau 1992: 94–5; Parfitt 2002).

45 Given the strong emphasis he places on the rules and discourses of disciplinary practice, Foucault's archaeological method is often (and erroneously, in his view) associated with structuralism. (Indeed, Morrow has called him a 'structuralist post-structuralist.') To distinguish himself from the structuralist school, Foucault explains in the Foreword to the English edition of *The Order of Things* (1966 [1970: xiv]) that:

> In France, certain half-witted 'commentators' persist in labeling me a 'structuralist'. I have been unable to get it into their tiny minds that I have used none of the methods, concepts, or key terms that characterize structural analysis. I should be grateful if a more serious public would free me from a connection that certainly does me honour, but that I have not deserved.

It is perhaps for this reason that Foucault moves away from the 'archaeological method,' in which he aims to develop 'a (structuralist) theory of rule-governed discursive practices,' in favour of a 'genealogy' that aims to document and understand the 'historical descent' of discursive practices, and the effect of discourses on human practice (Dreyfus and Rabinow 1983: xxv; Flyvbjerg 2001: 111; Parfitt 2002: 47).

46 The historical linkages between the international student protests of the late 1960s and the rise of 'new' intellectual and related social movements (especially women's liberation, black power, gay rights and, more generally, the politics of identity) are well considered in the excellent documentary on 'The Sixties,' by the Public Broadcasting System (PBS 2005).

47 Writing not long after this time, Laclau and Mouffe [(1985) 2001: 2] suggested that 'the very wealth and plurality of contemporary social struggles has given rise to a theoretical crisis.'

48 In so doing, he suggested that a distinction needs to be made between 'practical consciousness,' which implied the 'tacit stocks of knowledge which actors draw upon in the constitution of social activity,' and 'discursive consciousness, involving knowledge which actors are able to express on the level of discourse' (Giddens, 1979: 5).

49 Indeed, after presenting an earlier version of this chapter at the annual meeting of the International Studies Association in 2006, a fellow panelist remarked to me that his own interest in Marxism and dependency had long been stifled within his home department (of political science) at a large American university. He then went on to present what was effectively a regression analysis of multinational trade and investment in Latin America.

50 The contribution made by Amartya Sen we shall explore more carefully in Chapter Five.

51 Towards the end of his life, Foucault suggested in an interview with Paul Rabinow (cited in Parfitt 2002: 49–50) that insofar as it provided valuable social services, the welfare state served an essential function in modern society and, on this basis, he was unwilling to support policies or ideologies (including ones deriving from his own observations) that would undermine its existence. In its most 'benevolent' forms, the modern state reduced to 'docile bodies' the freedoms and traditions of the pre-modern era; however, the normalization of individual behaviour also provided tangible benefits that in the final analysis Foucault was unwilling to reject.

52 On university campuses in Europe and in North America, post-development 'readers,' such as Sachs (1992), Rahnema (1997), Schuurman (1993, 2000) and single-authored volumes by Ferguson (1990 [1994]) and Escobar (1995), became very popular texts for graduate and undergraduate courses in development, political science, anthropology and cultural studies. For an excellent overview of intellectual and historical linkages among postmodernism, post-colonialism and post-development, see Goss (1996), Sylvester (1999) and, especially, Chapter 5 by Richard Peet and Elaine Hartwick (1999) and Parfitt (2002).

53 'Discourse,' for Escobar (1995: 216), 'is not the expression of thought; it is a practice, with conditions, rules, and historical transformations.'

54 Here it should be stressed that many of the original 'contributions' to the *Post-Development Reader* had little if anything to do with the idea of 'post-development.' Indeed one could make the case that by 'extracting' these texts from their original context and purpose, Rahnema and Bawtree (1997) are in fact committing the same kinds of 'epistemic violence' that so infuriated the postmodernists (see Chapter Three). Although the authors claim in the Acknowledgments that they obtained permission from all of the authors (including Gandhi, whose death preceded the publication date by almost 50 years), and although some of the chapters (i.e. ones by Latouche, Illich, Shanin and Kothari) were original contributions, it is difficult to ignore the liberties the editors take in terms of associating this very wide assortment of essays and perspectives with their own agenda. Perhaps the most egregious example of this kind is the suggestion that the (liberal) political theorist Michael Ignatieff 'contributed' to the volume (by including his name on the back cover) when in fact his only 'contribution' can be found in two 'text boxes' (on pages 186 and 358). The editors defend this practice by suggesting that these and other text boxes are actually 'messages from absent friends or teachers who were either too far away or too busy to spend more time around the bigger table where the main conversation was being held' (Rahnema 1997a: xix). Leaving aside questions about what this conversation included and entailed, it is difficult to accept that the work of Ignatieff (whose writing is theoretically and ideologically very far from what is being advanced in the reader) can be construed in any way, shape or form as an argument in defense of vernacular society or simple living.

55 Partly for this reason, the line between post-developmental discourses (which questioned the achievements of development) and anti-development discourses (which rejected them altogether) became very fine indeed (cf. Corbridge 1998).

56 For Parfitt (2002), the relativism of post-development reflects the heavy reliance of post-development scholars (especially Escobar 1995; and Esteva and Prakash 1998) on the *early work* of Foucault, whose relativism he also subjects to critical scrutiny (Parfitt 2002: 50–2).

57 Indeed, more radical post-modernists would reject *prima facie* any effort to connect interpretation and identity with other forms of social and political action. For a thoughtful treatment of the ways in which relativist thinking in post-modernism undermines agency and coordinated action, see Rosenau (1992) and Parfitt (2002).

58 In a sense, Ferguson's treatment of development practice and failure bears strong similarity to what we might call 'conventional' political economy. In the case of the Thaba-Tseka livestock rehabilitation project, for instance, the project and the 'development problematic' failed because they were unable to 'apprehend' the economic structure of the local economy; and they 'failed to appreciate the larger political-economic situation of the project itself as an instrument of the government of Lesotho' (Ferguson 1990 [1994: 193]). In other words, the project failed to recognize the political structures and economic interests that governed the local economy (for livestock) and the relationship between society and the state. However, Ferguson is interested not (only) in the material factors that explain the failure of development interventions, but also in the ways in which 'failure' is construed and re-constructed in the face of these and other economic realities.

59 Indeed, Ferguson makes no mention of either the Northeast Asian experience or the (related) developmental state literature (e.g. Amsden, 1989; Johnson, 1982; Wade, 1990) which by the time of writing was a well-documented fact. For a good empirical treatment of this question, see Sandbrook *et al.* (2007), Kohli (2004), and Harriss (2001).

60 Indeed, the comparison can be taken further. Like Escobar (1995) and Esteva (1992), Easterly (2006: 24–5) highlights the construction of development and of development expertise, especially after Truman's Point Four speech in 1949. Unsurprisingly, Easterly (2006: 193–5) cites Ferguson (1990 [1994]) very favourably, primarily as a means of highlighting what he suggests is a pervasive lack of effectiveness and accountability in the foreign aid industry.

61 And here it is important to add that DFID is a relatively progressive institution, whose funding supported large numbers of critical social research projects, especially after the Labour Party's victory in 1997. For a particularly relevant example of DFID-supported programming resulting from critical social science research, see the *Postscript* by Corbridge *et al.* (2005: Chapter 9).

62 On this question, see Mosse (2005), Corbridge *et al.* (2005). As Corbridge *et al* (2005: 258–64) have pointed out, Ferguson appears to over-state the impact of a single World Bank country report on the actions and motivations of multiple development agencies and actors in the context of Lesotho.

63 However, as Sen points out (1999 [2001: 66]), this exemption was 'not very well integrated with the rest of Nozick's approach,' and therefore remained 'quite ad hoc.'

64 As Sen (1999 [2000]) has argued, 'traditional' (i.e. pre-quantitative) approaches aim to establish the value of economic decisions and social arrangements by assessing their ability to satisfy (through free market mechanisms) the maximum aggregate of human needs and desires. 'Modern' approaches, on the other hand, aim to rank individual preferences through the stated 'utility functions' (i.e. the stated utility derived from each consumption decision) drawn from a particular good or service.

65 'Functioning vectors' are simply a numerical representation of what a person may want to do or become in his or her life. One's 'capability set,' Sen (1999 [2001: 75]) argues, is a numerical representation of the opportunities (or alternatives) that a person has in his or her life, which differs from the 'realized functionings' of 'what a person is actually able to do.' The difference is roughly comparable to the distinction he makes between 'endowments' and 'entitlements'.

66 In this way, Sen's treatment of poverty is very similar to Robert Chambers' model of poverty spirals.

67 For a critical treatment of Sen's conceptualization of entitlement see Gore (1993), Gasper (1993) and Devereux (1996). Gore (1993: 442–3) and Devereux (1996: 4–5) argue that Sen's earlier work over-emphasizes the importance of state-recognized and state-enforced forms of entitlement. Responding to criticisms of this kind, Sen later expanded his conceptualization to include the idea of 'extended entitlement,' which includes customs, taboos and other informal norms and practices that determine questions of ownership, access and exchange in market settings.

68 Sen (1981) fully recognizes that the distribution of essential grains was not entirely the result of free market forces; considerable food was re-directed by colonial authorities during the famine to Calcutta and to other urban centres. However, the key point is that large segments of the population (principally those normally engaged in fishing and in agricultural labour) were excluded from the public distribution system and were therefore forced to establish entitlement over food and other essential commodities purely on the basis of their ability to sell labour (and other endowments), whose value in relation to the price of food had dropped significantly.

69 Individual liberties and well-being are part of the equation, but not in the way envisioned by libertarian and liberal theory. Particularly important for Sen is the Rawlsian 'priority of liberty' given to those 'classes of rights' which take precedence over any other social goal, 'including the removal of deprivation and destitution' (Sen 1999 [2001: 63]). Comparing the more 'demanding' libertarianism of Robert Nozick with the liberalism of John Rawls, Sen questions the moral basis upon which personal liberties (defined variously) could or should be elevated over and above 'the status of intense economic needs' (Sen 1999 [2001: 64]). For Sen, Nozick's assertion that individual freedoms (especially ones governing rights of private property) must be defended in the face of (almost) any conceivable consequence violates (or has the prospect of violating) 'the substantive freedom of individuals to achieve those things to which they have reason to attach great importance, including escaping avoidable mortality, being well-nourished and healthy, being able to read and count and so on.'

70 In so doing, he demonstrates persuasively that capability affects intensity of need and the ability to exploit opportunities in a free market economy.

71 Sen (1992: 55) broaches this subject (*inter alia*) when he reflects upon the 'non-grumbling resignation' of those facing long-standing inequality and deprivation. The argument he makes is that people may 'resign themselves' to chronic poverty and deprivation, and may therefore have no expectation of valuing a life beyond this experience. However, as Hilary Putnam (2002: 59) has argued, the recognition that values may vary does not resolve the problem of establishing consensus about 'which functionings are "valuable"' to which people. Moreover, as Des Gasper (2000: 999) puts it, 'Whether to have more options is valuable depends on the meaning the options have for the actor and her audience.'

72 Unlike Rawls, however, Sen's theory of justice is rooted not in a hypothetical condition, but in a proposition – based on empirical evidence and models of individual decision making and structured choice – about the need for individual responsibility and social commitment.

73 Indeed, a large proportion of *Development as Freedom* can be read as an effort to influence through the power of logic, evidence and reason the decisions of economists, politicians and other policy-making elites (cf. Gasper 2000).

74 Much has been made of Sen's interpretation of democracy in India, and the reference he makes to the 'pro-poor' agitations that led to the fall of Indira Gandhi's Congress coalition in 1977. However, his argument that the election 'turned' on the popular mobilization of 'the poor' dramatically simplifies important structural transformations that were ongoing at the time (e.g. class transformations arising as a result of the green revolution, the bifurcation of the Congress Party, the Emergency of 1975, etc.). For an excellent treatment of these questions and of this period, see Corbridge and Harriss (2000) and, especially, Corbridge *et al.* (2005: Chapter 2).

75 Similarly, his conceptualization of markets (Sen 1999 [2001]) is one that acknowledges clearly and explicitly the fact that labour is extracted, commodities traded and profits earned not only as a result of supply and demand, but also as a result of power, collusion, malfeasance, etc. Indeed, his treatment of freedom and markets (in Sen 1999 [2001: Chapter 5]) resonates strongly with the new institutional treatments of Douglass North (1990) and others.

76 Again the political economy literature on this question is large, and illuminating. For an excellent treatment of the politics of economic reform in the Indian context, see especially Jenkins (1999) and Corbridge and Harriss (2000).

77 In other words, consequences matter. For Sen, the formal articulation of rights and freedoms appears to take precedence over their actualization. Sen (1999 [2001: 230–1]) develops this argument by making an important distinction between rights that are not realized (in practice) and ones that do not exist.

78 Sen (1999 [2001: 10]) identifies five: political freedom, economic facilities, social opportunities, transparency guarantees and protective security.

79 And the challenge is large; Hilary Putnam (1981, cited in Kanbur and Shaffer 2006: 187), for instance, uses the sentence 'the cat is on the mat' to illustrate the large number of ontological and metaphysical assumptions we use to understand and communicate (apparently) simple concepts relating to animals, non-animals, spatial relations (being 'on' the mat), animation ('being'), etc. Although Putnam (2002: 13) also suggests that certain categories and concepts (e.g. string theory) may be more or less 'trivial' and complex than others (e.g. a cat on a mat), the analogy helps to illustrate the unspoken and unconscious assumptions we use to guide our communication with others.

80 Although he made no explicit reference to the dependency literature, Chambers in his classic *Rural Development: Putting the First Last* (1983: 4–5) in fact drew upon a simplified dependency model to highlight a 'core and periphery of knowledge' in which a rich, urban and industrialized elite draws 'resources and people away from the peripheries' and into the centres of national and international power and commerce. However, unlike much of the dependency literature, Chambers' critique (1983) also

entailed an explicit methodology, which he felt would provide both a workable means of understanding poverty and with it a means of intervening in ways that would alleviate the suffering of the poor.

81 Although he made no reference to the work of Foucault, Chambers quite clearly embraced the idea that development was constrained by the language and assumptions it used to understand poverty and the poor. Like Escobar (1995), he suggested that development research and practice rendered invisible certain categories of experience and people, especially ones in remote rural areas. However, unlike Escobar – and certainly unlike Foucault – Chambers (1983, 1997) also advanced an explicit way in which the personal and professional biases of development researchers and professionals could be reduced, if not eliminated.

82 The idea was a powerful one, and one that would influence subsequent trends in aid policy, especially the participatory ethos of the Poverty Reduction Strategies of the World Bank and bi-lateral donors, such as DFID. To understand the roots of this transition, see Killick *et al.* (1998). For an excellent treatment of PRSPs and their relation to neo-liberalism and World Bank policy, see Craig and Porter (2006).

83 The benefits of this approach were multiple. First, they respected the very large opportunity costs being asked of very poor people in the context of 'conventional' social science surveys. Second, the act of participating – and potentially humiliating oneself in the eyes of one's respondents – created a relationship that was more 'relaxed' and informal than the standard social science research exercise. Third, it gave researchers experiential insights about the particular nuances and demands of the livelihoods of the poor. For an excellent application of this methodological approach, see Leach and Fairhead (1996).

84 When done well, 'targeted' focus group discussions can isolate and ascertain the views and experiences of important and/or vulnerable groups in a community. As many critics (e.g. Kapoor 2002; Mosse 2001; Kothari 2001; Mohan and Stokke 2000) have argued, however, achieving these conditions – and overcoming power inequalities within communities – is difficult.

85 For an excellent summary and analysis of the SLA and of its intellectual history, see Solesbury (2003).

86 Although it recognizes the existence of power and the possibility of conflict, Kapoor argues, PRA fails to make explicit the principles upon which material conflicts of interest and ideological/normative differences would be reconciled or addressed. PRA lacks what he calls an 'adjudicating principle,' and resorts instead to the empiricist idea that differences may be accommodated and realities captured by developing 'more and better techniques' (Kapoor 2002: 108).

87 Indeed, for many 'specialists' (e.g. agronomists, hydrologists, Marxists) the SLA generates data whose connection to theory and policy is questionable and whose cumulative impact downgrades and sidelines professionally traditional areas of expertise (e.g. soil science, livestock management, etc.) in development practice (Sellamna 1999).

88 Much like the decentralization literature, ideas about how and why different forms of service delivery would work in different economic and political contexts have been influenced by 'policy experiments' carried out originally in industrialised countries. Now strongly touted by the World Bank (2004), ideas about voucher schemes, for instance, were based upon the experience of American cities (originally Milwaukee) which sought to give low-income families vouchers to enrol their children in private schools (Stein 2001: Chapter 3).

89 Although they clearly go beyond the scope of this chapter, explanations for the shift in thinking about rights and citizenship would include a growing sense of distrust towards government (Stein 2001), a perceived shift in taxation from corporations to individual incomes and purchases, a rise in civil and class action litigation, especially in the United States (Stein 2001; World Bank 2004) and finally, the spread of a 'consumer culture' in which the customer – or citizen – is 'always right' (Stein 2001).

90 For Booth (1993), the challenge of 'doing field research' in more than one location and the 'innate intellectual proclivities of those working at the micro, or the more actor-oriented end of the social research spectrum,' would produce a situation in which 'synthetic and comparative work (would lag) well behind the production of detailed empirical studies' (Booth 1993: 59).

91 Although they certainly cite him favourably, King *et al.* (1994) support an epistemology that appears quite different from Popper's. For King *et al.* (1994), induction and the interpretation of relevant facts and important debates constitute an important first step in the acquisition of knowledge.

92 For better or worse, *Designing Social Inquiry* has become one of the most influential texts in American social (or at least political) science (Johnson 2006). It has also generated considerable controversy and debate concerning *inter alia* its conceptualization of inference (e.g. Brady and Collier 2004; George and Bennett 2004; Johnson 2006), its characterization of 'qualitative research' (George and Bennett 2004; Johnson 2006) and its assertion that all social science inquiry should conform to the logic of inference. Some of these criticisms are explored below. For a more thorough treatment, see George and Bennett (2004), Johnson (2006) and Shapiro (2005).

93 Unlike Eckstein, King *et al.* (1994) favour the use of 'observation' because it avoids the confusion of a 'case' being equated with a 'case study,' which can include an infinite number of observations.

94 To be fair, the authors acknowledge early on in their work that they are dealing primarily with knowledge systems concerned with empirical inference, and that they quite consciously 'sidestep' more controversial questions relating to 'the nature and existence of truth, relativism and related subjects' (King *et al.* 1994: 6). That having been said, it is difficult to accept these assertions without questioning the processes by which importance, relevance and 'the real world' are defined in a scholarly context (Johnson 2006).

95 Indeed, unlike the behaviouralists, who went to great lengths to defend the separation of facts and values, their inherent skepticism about normative statements and the empirical study of human behaviour (see especially, Ricci 1984), the combined approaches being advanced by Laitin (2003) and King *et al.* (1994) appear to invest relatively little energy in this seemingly crucial exercise.

96 For dissenting and complementary perspectives on Tilly (1984), see the review essays by Wallerstein (1986) and Skocpol (1987), respectively.

97 It is worth noting here that India's federal system confers upon its states important powers governing, *inter alia*, agriculture, education, land distribution, labour and employment.

Bibliography

Acheson J. M. (1988) *The Lobster Gangs of Maine*. University Press of New England; London.

Agrawal A. (1994) Rules, rule making, and rule breaking: examining the fit between rule systems and resource use. In: (E. Ostrom, R. Gardner and J. Walker, eds) *Rules, Games and Common-Pool Resources*. University of Michigan Press; Ann Arbor, MI: pp. 267–82.

Agrawal A. (1999) *Greener Pastures: Politics, Markets and Community among a Migrant Pastoral People*. Oxford University Press; Delhi.

Agrawal A. (2001) Common property institutions and sustainable governance of resources. *World Development*; 29(10): 1649–72.

Agrawal A. (2002) Common resources and institutional stability. In: (E. Ostrom, ed) *The Drama of the Commons*. National Academy Press; Washington, DC: pp. 41–85.

Agrawal A. (2005) *Environmentality: Technologies of Government and the Making of Subjects*. Duke University Press; Durham, NC.

Agrawal A. and Yadama G. N. (1997) How do local institutions mediate market and population pressures on resources? Forest Panchayats in Kumaon, India. *Development and Change*; 28: 435–65.

Agrawal A and Gibson C. (1999) Enchantment and disenchantment: the role of community in natural resource conservation. *World Development*; 27 (4): 629–49.

Althusser L. (1969) *For Marx*. Translated by Ben Brewster. Penguin Press; London.

Alvarez W. and F. Asaro (1990) An Extraterrestrial Impact, *Scientific American* (October), 78–84.

Amin S. (1974) *Accumulation on a World Scale: A Critique of the Theory of Underdevelopment*. Monthly Review Press; New York.

Amsden A. (1989) *Asia's Next Giant: South Korea and Late Industrialization*. Oxford University Press; Oxford.

Angeles L. (2004) New issues, new perspectives: implications for development studies. *Canadian Journal of Development Studies*; XXV (1).

Badiou A. (2008) The Communist Hypothesis. *New Left Review*; (49) Jan/Feb: 29–42.

Baland J.-M. and Platteau J.-P. (1996). *Halting Degradation of Natural Resources: Is there a Role for Rural Communities?* Clarendon Press for the Food and Agriculture Organization; Oxford.

Baland J.-M. and Platteau J.-P. (1999) The ambiguous impact of inequality on local resource management. *World Development*; 27 (5): 773–88.

Ball T. and Dagger R. (2004) *Political Ideologies and the Democratic Ideal* (5th Edition) Pearson Longman; Toronto, ON.

Barrett C., Holden S. and Clay D. (2004) Can food-for-work programs reduce vulnerability? In: (S. Dercon, ed.) *Insurance against Poverty*. Oxford University Press; Oxford: pp. 361–84.

Barthes R. (1977) *Image-Music-Text.* Hill and Wang; New York.

Bates R. H. (1988) Contra contractarianism: some reflections on the new institutionalism. *Politics and Society*; 16 (2–3): 387–401.

Bates R. H. (1989) *Beyond the Miracle of the Market: The Political Economy of Agrarian Development in Kenya.* Cambridge University Press; Cambridge.

Bates R. H. (1995) Social Dilemmas and Rational Individuals: An Assessment of the New Institutionalism. In: Harriss J., Hunter J. and C. M. Lewis (eds) (1995) *The New Institutional Economics and Third World Development.* Routledge; London, pp. 27–48.

Bates R. H., Grief A., Levi M., Rosenthal J.-L. and Weingast B. (1998) *Analytic Narratives.* Princeton University Press; Princeton, NJ.

Bebbington A. (1999) Capitals and capabilities: a framework for analyzing peasant viability, rural livelihoods and poverty. *World Development*; 27 (12): 2021–44.

Bebbington A., Guggenheim S., Olson E. and Woolcock M. (2004) Exploring social capital debates at the World Bank. *Journal of Development Studies*; 40 (5): 33–64.

Beck T. (1994) *The Experience of Poverty: Fighting for Resources and Respect in Village India.* IT Publications; London.

Beck T. and Nesmith C. (2000) Building of poor people's capacities: the case of common property resources in India and West Africa. *World Development*; 29 (1): 119–33.

Beck U. (1992) *Risk Society: Towards a New Modernity.* Sage; London.

Bellamy Foster J. (1997) In defense of history. In: (E. Meiksins Wood and J. Bellamy Foster, eds) *In Defense of History: Marxism and the Postmodern Agenda.* Monthly Review Press; New York: pp. 184–94.

Bernstein H. (1983) Notes on capital and the peasantry. In: (J. Harriss, ed) *Rural Development: Theories of Peasant Economy and Agrarian Change.* Hutchinson University Library; London: pp. 160–77.

Bernstein, H. (1994) Agrarian classes in capitalist development. In: (L. Sklair, ed) *Capitalism and Development.* Routledge; London and New York: pp. 40–71

Bernstein H. (2005) 'Development Studies and the Marxists'. In: (U. Kothari, ed) *A Radical History of Development Studies: Individuals, Institutions and Ideologies.* London and New York; Zed Books, pp. 111–37.

Berry S. (1989) Social institutions and access to resources. *Africa* 59 (1): 41–55.

Bevan P. (2004) Studying poverty and inequality in poor countries: getting to grips with structure. *Unpublished Mimeo.* Department of Economics and International Development; University of Bath.

Black J. K. (1999) *Development in Theory and Practice: Paradigms and Paradoxes.* (2nd Edition) Westview Press; Boulder, CO.

Blaikie P. (1989) Environment and access to resources in Africa. *Africa*; 59 (1): 18–40.

Blaikie P. and Brookfield H. (1987) *Land Degradation and Society.* Methuen; London.

Blaikie P., Cannon T., Davis I. and Wisner B. (1994) *At Risk: Natural Hazards, People's Vulnerability and Disasters.* Routledge; London.

Blair H. (1996) Democracy, equity and common property resource management in the Indian subcontinent. *Development and Change*; 27 (3): 475–500.

Blomquist W. (1994) Changing rules, changing games: evidence from groundwater systems in Southern California. In: (E. Ostrom, R. Gardner and J. Walker, eds) *Rules, Games and Common-Pool Resources.* The University of Michigan Press; Ann Arbor, MI: pp. 283–300.

Blomquist W., Schlaeger E., Tang S. Y. and Ostrom E. (1994) Regularities from the field and possible explanations. In: (E. Ostrom, R. Gardner and J. Walker, eds) *Rules, Games and Common-Pool Resources.* The University of Michigan Press; Ann Arbor, MI: pp. 301–18.

Booth D. (1985) Marxism and development sociology: interpreting the impasse. *World Development*; 13 (7): 761–87.

Booth D. (1993) Development research: from 'impasse' to new development agenda. In: (F. Schuurman, ed.) (1993) *Beyond the impasse: new directions in development theory.* New York: Zed Books, pp. 49–76.

Booth D. (Ed.) (1994) *Rethinking social development: theory, research and practice.* Longman; London.

Booth D. (2003) 'Introduction and Overview', *Development Policy Review;* (21)2: pp. 131–159.

Bourdieu P. (1977) *Outline of a Theory of Practice.* Cambridge University Press; Cambridge.

Bracking S. (2004) Neoclassical and structural analysis of poverty: winning the 'economic kingdom' for the poor in Africa. *Third World Quarterly;* 25 (5): 887–901.

Bracking S. (2005) Guided miscreants: liberalism, myopias, and the politics of representation. *World Development;* 33 (6): 1011–24.

Brass T. (1991) Moral economists, subalterns, new social movements, and the (re-) emergence of a (post-) modernised middle peasant. *Journal of Peasant Studies;* 18 (2): 173–205.

Brenner R. (1977) 'The origins of capitalist development: a critique of neo-Smithian Marxism.' *New Left Review;* (104): July-August, 25–92.

Brett E. A. (1993) Voluntary agencies as development organizations: theorizing the problem of efficiency and accountability.' *Development and Change;* (24): 269–303.

Brett E. A. (2000) Understanding organizations and institutions. In: (D. Robinson, T. Hewitt and J. Harriss, eds) *Managing Development: Understanding Inter-Organizational Relationships.* Sage for the Open University Press; London: pp. 17–48.

Brigg M. (2002) Post-development, Foucault and the colonization metaphor. *Third World Quarterly;* 23 (3): 421–36.

Briggs J. and Sharp J. (2004) Indigenous knowledge and development: a postcolonial caution. *Third World Quarterly;* 25 (4): 661–76.

Brohmann J. (1995) Economism and critical silences in development studies: a theoretical critique of neoliberalism. *Third World Quarterly;* 16 (2): 297–318.

Bromley D. (1992) The commons, property and common-property regimes. In: (D. Bromley *et al.*, eds) *Making the Commons Work: Theory, Practice and Policy.* Institute for Contemporary Studies; San Francisco: pp. 3–16.

Bromley D. *et al.* (eds) (1992) *Making the Commons Work: Theory, Practice and Policy.* Institute for Contemporary Studies; San Francisco, CA.

Buttel F. and McMichael P. (1994) Reconsidering the explanandum and scope of development studies: toward a comparative sociology of state-economy relations. In: (D. Booth, ed) (1994) *Rethinking social development: theory, research and practice.* Longman; London: pp. 42–61.

Byres T. J. (1981) The new technology, class formation and class action in the countryside. *Journal of Peasant Studies;* 8 (4): 405–54.

Byres T. J. (1983) Agrarian transition and the agrarian question. In: (J. Harriss, ed) *Rural Development: Theories of Peasant Economy and Agrarian Change.* Hutchinson University Library; London: pp. 82–93.

Byres T. J. (1991) The agrarian question and differing forms of capitalist agrarian transition: an essay with reference to Asia. In: (J. Breman and S. Mundle, eds) *Rural Transformation in Asia.* Oxford University Press; Oxford: pp. 3–76.

Cameron J. (2005) Journeying in radical development studies: a reflection on thirty years of researching pro-poor development. In: (U. Kothari, ed) *A Radical History of Development Studies: Individuals, Institutions and Ideologies.* Zed Books; London and New York: pp. 138–56.

Campbell B., Mandondo A., Nemarundwe N., de Jong W., Luckert M. and Matose F. (2001) Challenges to proponents of common property resource systems: despairing voices from the social forests of Zimbabwe. *World Development;* 29 (4): 589–600.

Campbell J. L. and Pedersen O. K. (2001a) Introduction: the rise of neoliberalism and institutional analysis. In: (J. L. Campbell and O.K. Pedersen, eds) *The Rise of Neoliberalism and Institutional Analysis.* Princeton University Press; Princeton, NJ: pp. 1–24.

Campbell J. L. and Pedersen O.K. (2001b) The second movement in institutional analysis. In: (J. L. Campbell and O. K. Pedersen, eds) *The Rise of Neoliberalism and Institutional Analysis*. Princeton University Press; Princeton, NJ: pp. 249–82.

Cardoso F. H. (1979) *Dependency and Development in Latin America*. University of California Press; Berkeley, CA.

Carney D. (1998) Implementing the sustainable rural livelihoods approach. In: (D. Carney, ed) *Sustainable Rural Livelihoods: What Difference Can We Make?* Department for International Development; London: pp. 3–23.

Carr E. H. (1951) *The New Society*. Beacon Press; Boston, MA.

Carter M. (1997) Intellectual openings and policy closures: disequilibria in contemporary development economics. In: (F. Cooper and R. Packard, eds) *International Development and the Social Sciences: Essays on the History and Politics of Knowledge*. University of California Press; Berkeley, CA: pp. 119–49.

Chambers R. (1983) *Rural Development: Putting the Last First*. Longman; London.

Chambers R. (1997) Responsible well-being – a personal agenda for development. *World Development*; 25 (11): 1743–54.

Chambers R. (2005) Critical reflections of a development nomad. In: (U. Kothari, ed) *A Radical History of Development Studies: Individuals, Institutions, and Idealogies*. Zed Books Ltd; London: pp. 67–87.

Chambers R. and Conway G. (1991) Sustainable rural livelihoods: concepts for the 21st century. *IDS Discussion Paper 296*. Institute of Development Studies; Brighton.

Chandhoke N. (2002). Individual and group rights: a view from India. In: (E. Sridharan, Z. Hasan and R. Sudarshan, eds) *India's Living Constitution: Ideas, Practices, Controversies*. Permanent Black Publishing; Delhi: pp. 207–41.

Chandrasekhar C. P. (1993) Agrarian change and occupational diversification: non-agricultural employment and rural development in West Bengal. *Journal of Peasant Studies*; 20 (2): 205–70.

Chang H.-J. (2002) *Kicking Away the Ladder: Development Strategy in Historical Perspective*. Anthem Press; London.

Chilcote R. H. (1994) *Theories of Comparative Politics: The Search for a Paradigm Reconsidered* (2nd Edition). Westview Press; Boulder, CO.

Cleaver F. (2000) Moral ecological rationality, institutions and the management of common property resources. *Development and Change*; 31 (2): 361–83.

Cohn J. (1999) Irrational exuberance: when did political science forget about politics? *The New Republic*; October 25.

Coleman, J. S. (1990) *Foundations of Social Theory*. Belknap Press of Harvard University Press; London.

Collingwood R. G. (1946 [1992]) *The Idea of History*. Oxford University Press; Oxford.

Cooper F. and Packard R. (1997) Introduction. In: (F. Cooper and R. Packard, eds) *International Development and the Social Sciences: Essays on the History and Politics of Knowledge*. University of California Press; Berkeley, CA: pp.1–44.

Cooper F. and Packard R. (eds) (1997) *International Development and the Social Sciences: Essays on the History and Politics of Knowledge*. University of California Press; Berkeley, CA.

Corbridge S. (1986) *Capitalist World Development: A Critique of Radical Development Geography*. Rowan and Littlefield; Lanham, MD.

Corbridge S. (1990) Post-marxism and development studies: beyond the impasse. *World Development*; 18 (5): 623–39.

Corbridge S. (1993) Ethics in development studies: the example of debt. In: (F. Schuurman, ed) (1993) *Beyond the Impasse: New Directions in Development Theory*. Zed Books; New York: pp. 123–39.

Corbridge S. (1994) Post-Marxism and post-colonialism: the needs and rights of distant strangers. In: (D. Booth, ed) (1994) *Rethinking Social Development: Theory, Research And Practice*. Longman; London: pp. 90–117.

Corbridge S. (1998) Beneath the pavement only soil: the poverty of post-development. *Journal of Development Studies*; 34 (6): 138–48.

Corbridge S. and Harriss J. (2001) *Reinventing India: Liberalization, Hindu Nationalism and Popular Democracy*. Polity Press; Cambridge.

Corbridge S., Williams G., Srivastava M. and Véron R. (2005) *Seeing the State: Governance and Governmentality in India*. Cambridge University Press; Cambridge.

Cowen M. P. and Shenton R. W. (1996) *Doctrines of Development*. Routledge; London and New York.

Craig, D. and D. Porter (2006) *Development Beyond Neoliberalism?: Governance, Poverty Reduction and Political Economy*. Routledge; London and New York.

Crush J. (1995) Imagining development. In: (J. Crush, ed) *Power of Development*. Routledge; London and New York: pp. 1–26.

Dahlman C. (1980) *The Open Field System and Beyond: A Property Rights Analysis of an Economic Institution*. Cambridge University Press; Cambridge.

Davies S. (1996) How will you cope without me? A review of the literature on gender and coping strategies. *Unpublished mimeo*. Institute of Development Studies; Brighton.

Deleuze G. and Guattari F. (1997) *Anti-Oedipus: Capitalism and Schizophrenia*. Viking Press; London.

Derrida J. (1966 [1978]) Structure, sign, and play in the discourse of the human sciences. In: (J. Derrida, ed) *Writing and Difference*. (Trans. Alan Bass). Routledge; London: pp. 278–94.

Deshingkar P., Johnson C. and Farrington J. (2005) State transfers to the poor and back: the case of the food for work program in India. *World Development*; 33 (4): 575–91.

Devereux S. (1996) Fuzzy entitlements and common property resources: struggles over rights to communal land in Namibia. *IDS Working Paper*; (44). Institute of Development Studies; Brighton.

Deyo F. (1987) 'Introduction'. In: (F. Deyo, ed) *The Political Economy of the New Asia Industrialism*. Cornell University Press; Ithaca: pp. 11–22.

DFID (1999) *Sustainable Livelihoods Guidance Sheets*. Department for International Development; London.

Diegues A. C. (1998) Social Movements and the Remaking of the Commons in the Brazilian Amazon. In: (M. Goldman, ed) *Privatizing Nature: Political Struggles for the Global Commons*. Pluto Press; London: pp.54–75.

Dietz T., Dolsak N., Ostrom E. and Stern P. (eds) (2002) *The Drama of the Commons*. National Academy Press; Washington, DC.

DiJohn J. and Putzel J. (2000) State capacity building, taxation and resource mobilisation in historical perspective. *Presented at the New Institutional Economics, Institutional Reform and Poverty Reduction Conference*. Development Studies Institute, London School of Economics and Political Science; 7–8 September.

Djurfeldt G. (1982) Classical discussions of capital and the peasantry. In: (J. Harriss, ed) *Rural Development: Theories of Peasant Economy and Agrarian Change*. Hutchinson University Library; London: pp. 139–59.

Dreyfus H. L. and Rabinow P. (1983) *Michel Foucault, beyond structuralism and hermeneutics*. University of Chicago Press; Chicago.

Dreze J (2006) 'National Employment Guarantee Inaction,' *The Hindu*, 12 September 2006.

Dreze J. and Sen A. (1989) *Hunger and Public Action*. Clarendon Press; Oxford.

Eagleton T. (1997) Where do postmodernists come from? In: (E. Meiksins Wood and J. Bellamy Foster, eds) *In Defense of History: Marxism and the Postmodern Agenda* Monthly Review Press; New York: 17–25.

Easterly W. (2006) *The White Man's Burden: Why the West's Efforts to Aid the Rest Have Done So Much Ill and So Little Good*. The Penguin Press; New York.

Eco U. (1989) *Foucault's Pendulum*. Ballantine Books; New York.

Edwards M. (1993) How relevant is development studies? In: (F. Schuurman, ed) (1993) *Beyond the Impasse: New Directions in Development Theory.* New York: Zed Books, pp. 77–92.

Edwards M. (2002) Is there a 'future positive' for development studies? *Journal of International Development*; 14 (6): 737–41.

Edwards M. (2006) Looking back from 2046: 'Thoughts on the 80th anniversary of the Institute for Revolutionary Social Science.' Keynote speech to the 40th Anniversary of the Institute of Development Studies, University of Sussex, UK; downloaded at www.ids.ac.uk.

Ellis F. (1998) Livelihood diversification and sustainable rural livelihoods. In: (D. Carney, ed) *Sustainable Rural Livelihoods: What Difference Can We Make?* Department for International Development; London: pp. 53–66.

Emmanuel A. (1972) *Unequal Exchange: A Study of the Imperialism of Trade.* Monthly Review Press; New York.

Escobar A. (1995) *Encountering Development: The Making and Unmaking of the Third World.* Princeton University Press; Princeton, NJ.

Esteva G. (1992) Development. In: (W. Sachs, ed) *The Development Dictionary: A Guide to Knowledge and Power.* Zed Books; London: pp. 6–25.

Esteva G. and Prakash M. S. (1998) *Grassroots Postmodernism: Remaking the Soil of Cultures.* Zed Books; London.

Evans P. (1995) *Embedded Autonomy: States and Industrial Transformation.* Princeton University Press; Princeton.

Eyben R. (2006) The road not taken: international aid's choice of Copenhagen over Beijing. *Third World Quarterly*; 27 (4): 595–608.

Fairhead J. and Leach M. (1996) *Misreading the African Landscape: Society and Ecology in a Forest-Savanna Mosaic.* Cambridge University Press; Cambridge.

Fanon F. (1963) *The Wretched of the Earth.* Grove Press; New York.

Farmer P. (2004) An anthropology of structural violence. *Current Anthropology*; 45 (3): 305–24.

Farrington J. and Slater R. (2006) Cash transfers: panacea for poverty reduction or money down the drain? *Development Policy Review*; 24 (5): 499–511.

Ferguson J. (1990 [1994]) *The Anti-Politics Machine: 'Development,' Depoliticization and Bureaucratic Power in Lesotho.* University of Minnesota Press; Minneapolis, MN.

Ferguson J. (1997) Anthropology and its evil twin: 'development' in the constitution of a discipline. In: (F. Cooper and R. Packard, eds) *International Development and the Social Sciences: Essays on the History and Politics of Knowledge.* University of California Press; Berkeley, CA: pp. 150–75.

Fine B. (1999) The Developmental State is Dead – Long Live Social Capital? *Development and Change*; 30(1): 1–19.

Fine B. (2001) *Social Capital Versus Social Theory: Political Economy and Social Science at the Turn of the Millennium.* Routledge; London and New York.

Flyvbjerg B. (2001) *Making Social Science Matter: Why Social Inquiry Fails and How it Can Succeed Again.* Cambridge University Press; Cambridge.

Forsyth T. (2003) *Critical Political Ecology: The Politics of Environmental Science.* Routledge; New York and London.

Foucault M. (1961 [1965]) *Madness and Civilization: A History of Insanity in the Age of Reason.* Vintage Books; New York.

Foucault M. (1970) *The Order of Things.* Vintage Books; New York.

Foucault M. (1975 [1977]) *Discipline and Punish: The Birth of the Prison.* Vintage Books; New York.

Foucault M. (1978 [1990]) *The History of Sexuality: An Introduction Volume I.* Vintage Books; New York.

Fox J. (1996) How does civil society thicken? The political construction of social capital in rural Mexico. *World Development*; 24 (6): 1089–1103.

Frank A. G. (1969 [2000]) The development of underdevelopment. In: (J. T. Roberts and A. Hite, eds) *From Modernization to Globalization* Oxford. Blackwell; London: pp. 159–68.

Freedman R. (ed) (1961) *Marx on Economics.* Penguin Books; New York.

Frye N. (1958) *Three Lectures.* University of Toronto Press; Toronto, ON.

Galanter M. (2002) The long half-life of reservations. In: (Z. Hasan, E. Sridharan and R. Sudarshan, eds) *India's Living Constitution: Ideas, Practices, Controversies.* Permanent Black Publishing; Delhi: pp. 306–18.

Gandhi, M. (1997) 'The Quest for Simplicity: "My Idea of Swaraj",' in M. Rahnema with V. Bawtree (eds.) *The Post-Development Reader.* Zed Books; London: pp. 306–307.

Gasper D. (2000) 'Development as freedom: taking economics beyond commodities: the cautious boldness of Amartya Sen. *Journal of International Development*; 12 (7): 989–1001.

Gasper D. (2004) *The Ethics of Development.* Edinburgh University Press; Edinburgh.

Gauld R. (2000) Maintaining centralized control in community-based forestry: policy construction in the Philippines. *Development and Change*; 31 (1): 229–54.

Geertz C. (1973) *The Interpretation of Cultures.* Fontana Books; London.

George A. and Bennett A. (2004) *Case Studies and Theory Development in the Social Sciences.* MIT Press; Cambridge, MA.

Giddens A. (1971) *Capitalism and Modern Social Theory: An Analysis of the Writings of Marx, Durkheim and Max Weber.* Cambridge University Press; Cambridge.

Giddens A. (1979) *Central Problems in Social Theory: Action, Structure and Contradiction in Social Analysis.* Macmillan; London.

Giddens A. (1984) *The Construction of Society: Outline of the Theory of Structuration.* University of California Press; Berkeley, CA.

Gilpin, R. (2001) *Global Political Economy: Understanding the International Economic Order.* Princeton University Press; Princeton, NJ.

Goldman M. (ed.) (1998) *Privatizing Nature: Political Struggles for the Global Commons.* Rutgers University Press; New Brunswick, NJ.

Goldman M. (1998) Inventing the commons: theories and practices of the commons' professional. In: (M. Goldman, ed) *Privatizing Nature: Political Struggles for the Global Commons.* Rutgers University Press; New Brunswick, NJ: pp.20–53.

Gore C. (1993) Entitlement relations and 'unruly' social practices: a comment on the work of Amartya Sen. *Journal of Development Studies*; 29 (3): 429–60.

Gore C. (2000) The rise and fall of the Washington consensus as a paradigm for developing countries. *World Development*; 28 (5): 789–804.

Goss J. (1996) Postcolonialism: subverting whose empire? *Third World Quarterly*; 17 (2): 239–50.

Graaf J. (2006) The seductions of determinism in development theory: Foucault's functionalism. *Third World Quarterly*; 27 (8): 1387–1400.

Greenaway D. (1998) Does trade liberalisation promote economic development? *Southern Journal of Political Economy*; 45 (5): 491–511.

Habermas J. (1968 [1972]) *Knowledge and Human Interests.* Heinemann; London.

Habermas J. (1987) *The Philosophical Discourse of Modernity.* Cambridge University Press; Cambridge.

Hacking I. (1983) *Representing and Intervening: Introductory Topics in the Philosophy of Natural Science.* Cambridge University Press; Cambridge.

Hall P. A. and Taylor R. C. R. (1996) Political science and the three new institutionalisms. *Politics and Society*; 44 (5): 936–57.

Hancock G. (1989) *Lords of Poverty: The Power, Prestige and Corruption of the International Aid Business.* Atlantic Monthly Press; New York.

Hardin G. (1968 [2005]) The tragedy of the commons. In J. Dryzek and D. Schlosberg (eds) (2005) *Debating the Earth: The Environmental Politics Reader.* Second Edition. Oxford University Press; Oxford: pp. 25–36.

Harman C. (1982) *The Lost Revolution: Germany 1918 to 1923.* Bookmarks; London.

Harman C. (1987) *Explaining the Crisis: A Marxist Re-Appraisal.* Bookmarks; London.

Harris N. (1987) *The End of the Third World: Newly Industrializing Countries and the End of an Ideology.* Penguin; New York.

Harriss J. (1982) *Capitalism and Peasant Farming: Agrarian Structure and Ideology in Northern Tamil Nadu.* Oxford University Press; Oxford.

Harriss J. (1994) 'Between economism and post-modernism: reflections on the study of agrarian change in India'. In Booth, D (ed.) *Rethinking Social Development: Theory, Research and Practice.* Longman; Harlow: pp. 172–196.

Harriss J. (2000) How much difference does politics make? Regime differences across Indian states and rural poverty reduction. *Destin Working Paper No. 00–01*, Development Studies Institute, London School of Economics; London.

Harriss J. (2001) *Depoliticizing Development: The World Bank and Social Capital.* Anthem Press; London.

Harriss J. (2002) The case for cross-disciplinary approaches in international development. *World Development*; 30 (3): 487–96.

Harriss J. (2005) Great promise, hubris and recovery: a participant's history of development studies. In: (U. Kothari, ed) *A Radical History of Development Studies: Individuals, Institutions and Ideologies.* Zed Books; London and New York: pp. 17–46.

Harriss J. and De Renzio P. (1997) 'Missing link' or analytically missing? The concept of social capital. *Journal of International Development*; 9 (7): 919–37.

Harriss J., Hunter J. and Lewis C. M. (eds) (1995) *The New Institutional Economics and Third World Development.* Routledge; London.

Harriss J., Hunter J. and Lewis C. M. (1995a) Introduction: development and significance of NIE. In: (J. Harriss, J. Hunter and C. M. Lewis, eds) *The New Institutional Economics and Third World Development.* Routledge; London: pp.1–16.

Harriss-White B. (1996) Free market romanticism in an era of deregulation. *Oxford Development Studies*; 24 (1): 27–41.

Harvey D. (1990) *The Condition of Postmodernity.* Blackwell Press; Cambridge, MA.

Hayek, F. A. (1994 [1994]) The Road to Serfdom Chicago: University of Chicago Press.

Hettne B. (1995) *Development Theory and the Three Worlds: Towards an International Political Economy of Development.* Longman; Harlow, UK.

Hickey S. and Bracking S. (2005) Exploring the politics of chronic poverty: from representation to a politics of justice? *World Development*; 33 (6): 851–65.

Hickey S. and Mohan G. (2005) Relocating participation within a radical politics of development. *Development and Change*; 36 (2): 237–262.

Hindess B. and Hirst P. (1977) *Mode of Production and Social Formation: An Auto-Critique of 'Pre-Capitalist Modes of Production.* Macmillan Press; London.

Hirschmann A. (1970). *Exit, voice and loyalty: responses to decline in firms, organizations and states.* Harvard University Press; Cambridge, MA.

Hobsbawm E. (1987 [1989]) *The Age of Empire 1875–1914.* Vintage Books; New York.

Hodgson G. M. (1993) Institutional economics: surveying the 'old' and the 'new.' *Metroeconomica*; 44 (1): 1–28.

Hollis M. (1994) *The Philosophy of Social Science: An Introduction.* Cambridge University Press; Cambridge.

Homer-Dixon T. (1991) On the threshold: environmental changes as causes of acute conflict. *International Security*; 16 (2): 76–116.

Hoogvelt A. (2001) *Globalization and the Post-Colonial World: The New Political Economy of Development.* Johns Hopkins University Press; Baltimore, MA.

Huntington S. (1968) *Political Order in Changing Societies.* Yale University Press; New Haven, CT.

Ignatieff M. (2000) *The Rights Revolution.* House of Anansi Press; Toronto, ON.

Illich I. (1997) Development as planned poverty. In: (M. Rahnema and V. Bawtree, eds) *The Post-Development Reader.* Zed Books; London: pp. 94–102.

Immergut E. (1998) The theoretical core of the new institutionalism. *Politics and Society*; 26 (1): 5–34.

Isbister J. (2006) *Promises Not Kept: Poverty and the Betrayal of Third World Development* (7th Edition). Kumarian Press; West Hartford, CT.

Jackson C. (2002) Disciplining gender? *World Development*; 30 (3): 497–509.

Jameson F. (1984) Postmodernism, or the cultural logic of late capitalism. *New Left Review*; 146: 53–92.

Jenkins R. (1999) *Democratic Politics and Economic Reform in India.* Cambridge University Press; Cambridge.

Jodha N. S. (2001) *Life on the Edge: Sustaining Agriculture and Community Resources in Fragile Environments.* Oxford University Press; Delhi.

Johnson C. (1982) *MITI and the Japanese Miracle: The Growth of Industrial Policy, 1925–1975.* Stanford University Press; Stanford.

Johnson, C. (2000) *Common property, political economy and institutional change: community-based management of an inshore fishery in southern Thailand.* Unpublished Ph.D. London School of Economics; London.

Johnson C. (2001) Community formation and fisheries conservation in Southern Thailand. *Development and Change*; 32 (5): 951–74.

Johnson C. (2002) State and community in rural Thailand: Village society in historical perspective, *The Asia Pacific Journal of Anthropology*; 2(2): 114–134.

Johnson C. (2004) Uncommon ground: the 'poverty of history' in common property discourse. *Development and Change*; 35 (3): 407–33.

Johnson C. and Forsyth T. (2002) In the eyes of the state: negotiating a rights-based approach to forest conservation in Thailand. *World Development*; 30 (9): 1591–1605.

Johnson C., Deshingkar P. and Start D. (2005) Grounding the state: devolution and development in India's *panchayats. The Journal of Development Studies*; 41 (5): 937–70.

Johnson E. and Stone D. (2000) The Genesis of the GDN. In: (D. Stone, ed) *Banking on Knowledge: The Genesis of the Global Development Network.* Routledge; London and New York: pp. 3–23.

Johnson J. (2006) Consequences of positivism: a pragmatist assessment. *Comparative Political Studies*; 39 (2): 224–52.

Kabeer N. (1994) *Reversed Realities: Gender Hierarchies in Development Thought.* Verso; London.

Kanbur R. (2001) Economic policy, distribution and poverty: the nature of disagreements. *World Development*; 29 (6): 1083–94

Kanbur R. (2002) Economics, social science and development. *World Development*; 30 (3): 477–86.

Kanbur R. and Shaffer P. (2006) Epistemology, normative theory and poverty analysis: implications for Q-squared in practice. *World Development*; 35 (2): 183–96.

Kapoor I. (2002) The devil's in the theory: a critical assessment of Robert Chambers' work on participatory development. *Third World Quarterly*; 23 (1): 101–17.

Kapoor I. (2004) Hyper-self-reflexive development? Spivak on representing the Third World 'other.' *Third World Quarterly*; 25 (4): 627–47.

Keohane R. O. and Ostrom E. (eds) (1995) *Local Commons and Global Interdependence: Heterogeneity and Cooperation in Two Domains.* Sage Publications; London.

Khilnani S. (2002) The Indian constitution and democracy. In: (Z. Hasan, E. Sridharan and R. Sudarshan, eds) *India's Living Constitution: Ideas, Practices, Controversies.* Permanent Black Publishing; Delhi: pp. 64–82.

Killick T. (1998) *Aid and the Political Economy of Policy Change.* Routledge; London and New York.

King G., Keohane R. O. and Verba S. (1994) *Designing Social Inquiry: Scientific Inference in Qualitative Research.* Princeton University Press; Princeton, NJ.

Knight J. (2001) Explaining the rise of neoliberalism: the mechanisms of institutional change. In: (J. L. Campbell and O. K. Pedersen, eds) *The Rise of Neoliberalism and Institutional Analysis.* Princeton University Press; Princeton: pp. 27–50.

Kohli A. (2004) *State-Directed Development: Political Power and Industrialization in the Global Periphery.* Cambridge University Press; Cambridge.

Kothari R. (1997) 'The Agony of the Modern State' in M. Rahnema with V. Bawtree (eds) *The Post-Development Reader.* Zed Books; London: pp. 143–151.

Kothari U. (2001) Power, knowledge and social control in participatory development. In: (B. Cooke and U. Kothari, eds) *Participation: The New Tyranny?* Zed Books; London: pp. 139–52.

Kothari U. (ed) (2005) *A Radical History of Development Studies: Individuals, Institutions and Ideologies.* Zed Books; London and New York.

Kothari U. (2005) From colonial administration to development studies: a post-colonial critique of the history of development studies. In: (U. Kothari, ed) *A Radical History of Development Studies: Individuals, Institutions and Ideologies.* Zed Books; London and New York: pp. 47–66.

Krueger A. (1998) Why trade liberalisation is good for economic growth. *The Economic Journal*; 108 (450): 1513–22.

Kuhn T. (1962) *The Structure of Scientific Revolutions.* University of Chicago Press; Chicago.

Kurien J. (1992) Ruining the commons and responses of the commoners: coastal overfishing and fishworkers' actions in Kerala state, India. In: (D. Ghai and J. Vivian, eds) *Grassroots Environmental Action: People's Participation in Sustainable Development.* Routledge; London: pp. 221–58.

Laclau E. and Mouffe C. (1985 [2001]) *Hegemony and Socialist Strategy: Towards a Radical Democratic Politics.* Verso; London.

Laitin D. D. (2003) The Perestroikan Challenge to Social Science. *Politics and Society*; 31 (1): 163–84.

Lal D. (1983 [2000]) *The Poverty of 'Development Economics.'* Oxford University Press; New York.

Lam W. F. (1996) Institutional design of public agencies and coproduction: a study of irrigation associations in Taiwan. *World Development*; 24 (6): 1039–54.

Landes D. (1998) *The Wealth and Poverty of Nations.* W.W. Norton; New York.

Lapham L. H. (2006) Notebook: blue guitar. *Harper's Magazine*; 312 (1872): 15.

Leach M., Mearns R. and Scoones I. (1999) Environmental entitlements: dynamics and institutions in community-based natural resource management. *World Development*; 27 (2): 225–47.

Lehmann D. (1997) An opportunity lost: Escobar's deconstruction of development. *Journal of Development Studies*; 33 (4): 568–78.

Lenin V. I. (1983) The differentiation of the peasantry. In: (J. Harriss, ed) *Rural Development: Theories of Peasant Economy and Agrarian Change.* Hutchinson University Library; London: pp. 130–8.

Lerner D. (1958 [2000]) The passing of traditional society. In: (J. T. Roberts and A. Hite, eds) *From Modernization to Globalization* Oxford: Blackwell, London: pp. 119–32.

Levi M. (1988). *Of Rule and Revenue.* University of California Press; Berkeley, LA.

Lewis D., Rodgers D and Woolcock M. (2008) The fiction of development: literary representation as a source of authoritative knowledge. *Journal of Development Studies*; 44 (2): 198–216.

Leys C. (1996) *The Rise and Fall of Development Theory.* Indiana University Press; Bloomington, IN.

Li T. M. (1996) Images of community: discourse and strategy in property relations. *Development and Change*; 27 (3): 501–27.

Libecap G. D. (1995) The conditions for successful collective action. In: (R. O. Keohane and E. Ostrom, eds) (1995) *Local Commons and Global Interdependence: Heterogeneity and Cooperation in Two Domains.* Sage Publications; London: pp. 161–90.

Long N. and Villarreal M. (1993) 'Exploring development interfaces: from the transfer of knowledge to the transfer of meaning. In: (F. Schuurman, ed) (1993) *Beyond the Impasse: New Directions in Development Theory.* Zed Books; New York: pp. 140–68.

Long N. and van der Ploeg J. D. (1994) Heterogeneity, actor and structure: towards a reconstitution of the concept of structure. In: (D. Booth, ed) (1994) *Rethinking Social Development; Theory, Research and Practice.* Longman; London: pp. 62–89.

McCay B. J. and Acheson J. M. (eds) (1987) *The Question of the Commons: The Culture and Ecology of Communal Resources.* University of Arizona Press; Tucson, AZ.

McCay B. J. and Jentoft S. (1998) Market or community failure? Critical perspectives on common property research. *Human Organization*; 57 (1): 21–9.

McMichael P. (2000) World-systems analysis, globalization and incorporated comparison. *Journal of World-Systems Research*; Fall/Winter VI (3): 68–99.

McMichael P. (2004) *Development and Social Change: A Global Perspective Third Edition.* Pine Forge Press; Thousand Oaks, CA.

Macpherson C. B. (1973) *Democratic Theory: Essays in Retrieval.* Oxford University Press; New York.

Macpherson C. B. (1978) *Property: Mainstream and Critical Positions.* University of Toronto Press; Toronto, ON.

Marx K. (1867 [1976]) *Capital Volume I.* Penguin; London.

Marx K. (1963) *The Eighteenth Brumaire of Louis Bonaparte.* International Publishers; New York.

Marx K. (1971) *A Contribution to the Critique of Political Economy.* Lawrence and Wishart; London.

Marx K. and Engels F. (1888 [1967]) *The Communist Manifesto.* Penguin Books; New York.

Marx K. and Engels F. (1964) *The German Ideology.* Progress Publishers; Moscow.

Mbembe A. (2003) Necropolitics. *Public Culture*; 15 (1): 11–40.

Meadows D., Meadows D., Randers J. and Behrens III W. W. (1972) *Limits to Growth.* Universe Books; New York.

Mearns R. (1996) Environmental entitlements: pastoral natural resource management in Mongolia. *Cahiers des Sciences Humaines*; 32 (1): 105–31.

Meinzen-Dick R., Raju K. V. and Gulati A. (2002) What affects organization and collective action for managing resources? Evidence from canal irrigation systems in India. *World Development*; 30 (4): 649–66.

Meiksins Wood E. (1997) What is the 'postmodern' agenda? In: (E. Meiksins Wood and J. Bellamy Foster, eds) *In Defense of History: Marxism and the Postmodern Agenda.* Monthly Review Press; New York: pp. 1–16.

Migdal J. (1988) *Strong Societies and Weak States: State-Society Relations and State Capabilities in the Third World.* Princeton University Press; Princeton, NJ.

Mitchell R. B. (1995) Heterogeneities at two levels: states, non-state actors and intentional oil pollution. In: (R. O. Keohane and E. Ostrom, eds) (1995) *Local Commons and Global Interdependence: Heterogeneity and Cooperation in Two Domains.* Sage Publications; London: pp. 223–52.

Mohan G. and Stokke K. (2000) Participatory development and empowerment: the dangers of localism. *Third World Quarterly*; 21 (2): 247–68.

Moore Jr B. (1966) *Social Origins of Dictatorship and Democracy.* Penguin Books; New York.

Moore M. and Putzel J. (1999) Politics and Poverty: A Background Paper for the World Development Report 2000/1. *Mimeo.* Institute of Development Studies; Brighton.

Morrow R. A. and Brown D. D. (1994) *Critical Theory and Methodology.* Sage; Thousand Oaks, CA.

Mosse D. (1997) The symbolic making of a common property resource: history, ecology and locality in a tank-irrigated landscape in South India. *Development and Change*; 28 (3): 467–504.

Mosse D. (2001) People's knowledge, participation and patronage: operations and representations in rural development. In: (B. Cooke and U. Kothari, eds) *Participation: The New Tyranny?* Zed Books; London: pp. 16–35.

Mosse D. (2005) *Cultivating Development: An Ethnography of Aid Policy and Practice* Pluto Press; London.

Munck R. (1999) Deconstructing development discourses: of impasses, alternatives and politics. In: (R. Munck and D. O'Hearn, eds) *Critical Development Theory: Contributions to a New Paradigm.* Zed Books; London: pp. 196–210.

Nandy A. (1997) Colonization of the mind. In: (M. Rahnema and V. Bawtree, eds) *The Post-Development Reader.* Zed Books; London: pp. 168–78.

Nederveen Pieterse J. (1998) My paradigm or yours? Alternative development, post-development, reflexive development. *Development and Change*; 29 (2): 343–73.

Nederveen Pieterse, J. (2000) After post-development. *Third World Quarterly*; 21 (2): 175–91.

Netting R. (1981) *Balancing on an Alp: Ecological Change and Continuity in a Swiss Mountain Community.* Cambridge University Press; Cambridge.

Nguiffo S.-A. (1998) In defence of the commons: forest battles in southern Cameroon. In: (M. Goldman, ed) (1998) *Privatizing Nature: Political Struggles for the Global Commons.* Rutgers University Press; New Brunswick, NJ: pp.102–19.

Norberg-Hodge H. (1997) Learning from Ladakh. In: (M. Rahnema and V. Bawtree, eds) *The Post-Development Reader.* Zed Books; London: pp. 22–9.

North D. C. (1990) *Institutions, Institutional Change and Economic Performance.* Cambridge University Press; Cambridge.

North D. C. (1995) The new institutional economics and Third World development. In: (J. Harriss, J. Hunter and C. M. Lewis, eds) (1995) *The New Institutional Economics and Third World Development.* Routledge; London: pp.17–26.

Nozick R. (1974) *Anarchy, State and Utopia.* Basic Books; New York.

Nussbaum M. (2000) *Women and Human Development: The Capabilities Approach* Cambridge University Press; Cambridge.

Nustad K. (2001) Development: the devil we know? *Third World Quarterly*; 22 (4): 479–89.

Olson M. (1965) *The Logic of Collective Action: Public Goods and the Theory of Groups.* Harvard Press; London.

Onis Z. and Senses F. (2005) Re-thinking the emerging post-Washington consensus. *Development and Change*; 36 (2): 263–90.

Ostrom E. (1990) *Governing the Commons: the Evolution of Institutions for Collective Action.* Cambridge University Press; Cambridge.

Ostrom E. (1991) Rational choice theory and institutional analysis: toward complementarity. *American Political Science Review*; 85 (1): 237–43.

Ostrom E. (1998) A behavioral approach to the rational choice theory of collective action: presidential address, American Political Science Association, 1997. *American Political Science Review*; 92 (1): 1–22.

Ostrom E. (2000) Private and common property rights. *Unpublished mimeo*, downloaded from the Library of the Commons, http://www.indiana.edu/~iascp/Iforms/searchcpr.html, accessed 7 May 2003.

Ostrom E., Gardner R. and Walker J. (1994) *Rules, Games and Common-Pool Resources.* University of Michigan Press; Ann Arbor, MI.

Ostrom E. Gardner R. and Walker J. (1994a) Cooperation and social capital. In: E. Ostrom, R. Gardner and J. Walker (1994) *Rules, Games and Common-Pool Resources.* University of Michigan Press; Ann Arbor, MI, pp. 319–29.

Ostrom E. Schroeder L. and Wynne S (1993) *Institutional Incentives and Sustainable Development: Infrastructure Policies in Perspective.* Westview Press; Oxford.

Page S. and Hewitt A. (2001) *World Commodity Prices: Still a Problem for Developing Countries?* Overseas Development Institute; London.

Parfitt T. (2002) *The End of Development: Modernity, Post-Modernity and Development.* Pluto Press; London.

Parfitt T. (2004) The ambiguity of participation: a qualified defence of participatory development. *Third World Quarterly*; 25 (3): 537–56.

Parpart J. and Veltemeyer H. (2004) The development project in theory: a review of its shifting dynamics. *Canadian Journal of Development Studies*; XXV (1): 39–60.

Parsons T. (1964 [2000]) Evolutionary universals in society. In: (J. T. Roberts and A. Hite, eds) *From Modernization to Globalization Oxford*: Blackwell; London: pp. 83–99.

PBS (2005) *The Sixties: The Years that Shaped the Generation.* The Public Broadcasting Service.

Pearce D. W. and Warford J. J. (1993) *World Without End.* Oxford University Press; Oxford.

Pearson R. (2005) The rise and rise of gender and development. In: (U. Kothari, ed) *A Radical History of Development Studies: Individuals, Institutions and Ideologies.* Zed Books; London and New York: pp. 157–79.

Peet R. (1991) *Global Capitalism: Theories of Societal Development.* Routledge Press; London and New York.

Peet R. (1994) Review of *Beyond the impasse: new directions in development theory. Annals of the Association of American Geographers*; 84 (2): 339–42.

Peet R. and Watts M. (eds) (1996) *Liberation Ecologies: Environment, Development, Social Movements.* Routledge Press; London and New York.

Peet R. and Hartwick E. (1999) *Theories of Development.* The Guilford Press; New York.

Peters B. G. (1998) *Comparative Politics: Theory and Methods.* New York University Press; New York.

Pierson P. (2004) *Politics in Time: History, Institutions and Social Analysis.* Princeton University Press; Princeton, NJ.

Polanyi K. (1957) *The Great Transformation: The Political and Economic Origins of Our Time.* Beacon Press; Boston, MA.

Pollard S. (1968) *The Idea of Progress: History and Society.* C. A. Watts; London.

Popkin S. (1979) *The Rational Peasant: The Political Economy of Rural Society in Vietnam.* University of California Press; Berkeley, CA.

Popper K. (1957 [1997]) *The Poverty of Historicism.* Routledge; London and New York.

Popper K. (1962) *Conjectures and Refutations: The Growth of Scientific Knowledge.* Basic Books; New York.

Prakash S. (1998) Fairness, social capital and the commons: the societal foundations of collective action in the Himalaya. In: (M.Goldman, ed) (1998) *Privatizing Nature: Political Struggles for the Global Commons.* Rutgers University Press; New Brunswick, NJ: pp. 167–97.

Putnam H. (2002) *The Collapse of the Fact-Value Dichotomy and Other Essays.* Harvard University Press; Cambridge, MA.

Putnam, R.D. (1993) *Making Democracy Work: Civic Traditions in Modern Italy.* Princeton University Press; Princeton, NJ.

Putzel J. (1992) *A Captive Land: The Politics of Agrarian Reform in the Philippines.* Catholic Institute for International Relations; London.

Putzel J. (1997) Accounting for the 'dark side' of social capital: reading Robert Putnam on democracy. *Journal of International Development*; 9 (7): 939–49.

Rahnema M. (1997a) Development and the people's immune system: the study of another variety of AIDS. In: (M. Rahnema and V. Bawtree, eds) *The Post-Development Reader.* Zed Books; London: pp. 111–34.

Rahnema M. (1997b) Towards post-development: searching for signposts, a new language and new paradigms. In: (M. Rahnema and V. Bawtree, eds) *The Post-Development Reader*; Zed Books; London: pp. 377–404.

Rahnema M. and V. Bawtree (eds) (1997) *The Post-Development Reader.* Zed Books; London.

Rapley J. (2002) *Understanding Development: Theory and Practice in the Third World.* (2nd Edition) Lynne Reinner; Boulder, CO.

Rawls J. (1971) *A Theory of Justice.* Belknap Press of Harvard University Press; Cambridge, MA.

Redden C. J. (2002). Health care as citizenship development: examining social rights and entitlement. *Canadian Journal of Political Science*; 35 (1): 103–25.

Rew A., Khan S. and Rew M. (2006) 'P3>Q2 in Northern Orissa: an example of integratig 'combined methods' (Q2) through a 'platform for probing poverties' (P3). *World Development*; 35 (2): 281–95.

Ribot J. (1998) Theorizing access: forest politics along Senegal's charcoal commodity chain. *Development and Change*; 29 (2): 307–41.

Ricci D. (1984) *The Tragedy of Political Science: Politics, Scholarship and Democracy.* Yale University Press; New Haven, CT.

Rist G. (1997) *The History of Development: From Western Origins to Global Faith.* Zed Books; New York.

Rosenau P. M. (1992) *Post-Modernism and the Social Sciences: Insights, Inroads and Intrusions.* Princeton University Press; Princeton, NJ.

Ross E. (1998) *The Malthus Factor: Poverty, Politics and Population in Capitalist Development.* Zed Books; New York.

Rossi B. (2004) Revisiting Foucauldian approaches: power dynamics in development projects. *Journal of Development Studies*; 40 (6): 1–29.

Rostow W. W. (1960) *The Stages of Economic Growth: A Non-Communist Manifesto.* Cambridge University Press; Cambridge.

Roxborough I. (1979) *Theories of Underdevelopment.* MacMillan; London.

Saberwal, S. (2002) 'Introduction: Civilization, constitution, democracy', in Z. Hasan et al. (eds) *India's Living Constitution: Ideas, Practices, Controversies.* Permanent Black Publishing; Delhi: pp. 1–30.

Sachs W. (1992) Introduction. In: (W. Sachs, ed) *The Development Dictionary: A Guide to Knowledge and Power.* Zed Books; London: pp. 1–5.

Sahlins M. (1997) 'The Original Affluent Society,' in M. Rahnema with V. Bawtree (eds) *The Post-Development Reader.* Zed Books; London: pp. 3–21.

Said E. (1979 [1994]) *Orientalism.* Vintage Books; New York.

Saith A. (2006) From universal values to millennium development goals: lost in translation. *Development and Change*; 37 (6): 1167–99.

Salemink O. (2003) 'Social science intervention: moral versus political economy and the Vietnam war.' In: (P. Quarles Van Ufford and A. K. Giri, eds) *A Moral Critique of Development: In Search of Global Responsibilities.* Routledge; London and New York: pp. 169–93.

Sandbrook R., Edelman M., Heller P. and Teichman J. (2007) *Social Democracy in the Global Periphery: Origins, Challenges, Prospects.* Cambridge University Press; Cambridge.

Sarup M. (1989) *An Introductory Guide to Post-Structuralism and Postmodernism.* University of Georgia Press; Athens, GA.

Schlager E. and Ostrom E. (1992) Property rights regimes and natural resources: a conceptual analysis. *Land Economics*; 68 (3): 249–62.

Schuurman F. (ed) (1993) *Beyond the Impasse: New Directions in Development Theory.* Zed Books; New York.

Schuurman F. (1993) Introduction: development theory in the 1990s. In: (F. Schuurman, ed) *Beyond the Impasse: New Directions in Development Theory.* Zed Books; New York: pp. 1–48.

Schuurman F. (2000) Paradigms lost, paradigms regained? Development studies in the twenty-first century. *Third World Quarterly*; 21 (1): 7–20.

Schuurman F. (2001) Globalization and development studies: Introducing the challenges. In: (F. Schuurman, ed) *Globalization and Development Studies: Challenges for the 21st Century.* Sage; London: pp. 3–19.

Scoones I. (1996) Ecological dynamics and grazing resource tenure: a case study from Zimbabwe. *Unpublished mimeo.* Institute of Development Studies; Brighton.

Scoones I. (1998) Sustainable rural livelihoods: a framework for analysis. *IDS Working Paper 72.* Institute of Development Studies; Brighton.

Scoones I. (1999) New ecology and the social sciences: what prospects for fruitful engagement? *Annual Review of Anthropology*; 28 (October): 479–507.

Scott J. C. (1976) *The Moral Economy of the Peasant: Rebellion and Subsistence in Southeast Asia.* Yale University Press; New Haven, CT.

Scott J. C. (1985) *Weapons of the Weak: Everyday Forms of Peasant Resistance.* Yale University Press; New Haven, CT.

Scott J. C. (1998) *Seeing Like a State: How Certain Schemes to Improve the Human Condition Have Failed.* Yale University Press; New Haven, CT.

Sellamna N.-E. (1999) Relativism in agricultural research and development: is participation a post-modern concept? *ODI Working Paper 119.* Overseas Development Institute; London.

Sen A. (1981) *Poverty and Famines.* Oxford University Press; New York and Oxford.

Sen A. (1985) *Commodities and Capabilities.* Elsevier; New York.

Sen A. (1992) *Inequality Re-examined.* Harvard University Press; Cambridge, MA.

Sen A. (1999 [2001]) *Development as Freedom.* Oxford University Press; New York and Oxford.

Shapiro I. (2005) *The Flight from Reality in the Human Sciences.* Princeton University Press; Princeton, NJ.

Shanin T. (1997) The idea of progress. In: (M. Rahnema and V. Bawtree, eds) *The Post-Development Reader.* Zed Books; London: pp. 65–71.

Shiva V. (1997) Western science and its destruction of local knowledge. In: (M. Rahnema and V. Bawtree, eds) *The Post-Development Reader.* Zed Books; London: pp. 161–7.

Simon D. (1997) Development reconsidered: new directions in development thinking. *Geografiska Annaler*; (79B): 183–201.

Simon D. (2002) Postmodernism and development. In: (V. Desai and R. Potter, eds) *The Companion to Development Studies.* Arnold; London: pp. 121–7.

Simon D. (2007) Beyond antidevelopment: discourses, convergences, practices. *Singapore Journal of Tropical Geography*; 28 (2): 205–18.

Simon J. (1981) *The Ultimate Resource.* Princeton University Press; Princeton, NJ.

Sinha S., Gururani S, and Greenberg B. (1997) The 'new traditionalist' discourse of Indian environmentalism. *Journal of Peasant Studies*; 24 (3): 65–99.

Sklair L. (1988) Transcending the impasse: metatheory, theory and empirical research in the sociology of development and underdevelopment. *World Development*; 16 (6): 697–709.

Skocpol T. (1979) *States and Social Revolutions: A Comparative Analysis of France, Russia and China.* Cambridge University Press; Cambridge.

Skocpol T. (1987) Social history and historical sociology: contrasts and complementarities. *Social Science History*; 11 (1): 17–30.

Slater D. (1993) The geopolitical imagination and the enframing of development theory. *Transactions. Institute of British Geographers. N. S.* (18): 419–437.

Slater D. (1997) Geopolitical imaginations across the north-south divide. *Political Geography*; 16 (8): 631–53.

Solesbury W. (2003) Sustainable livelihoods: a case study of the evolution of DFID policy. *ODI Working Paper 217.* Overseas Development Institute; London.

Spivak G. C. (1988) Can the subaltern speak?' In: (C. Nelson and L. Grossman, eds) *Marxism and the Interpretation of Culture.* University of Illinois Press; Urbana, IL: pp. 271–313.

Stedman Jones G. (2002) *Introduction to the Communist Manifesto.* Penguin Books; New York.

Stein J. G. (2001) *The Cult of Efficiency.* House of Anansi Press; Toronto, ON.

Stiglitz J. (2003) *Globalization and its Discontents.* Norton; New York.

Sumner A. (2007) Meaning versus measurement: why do economic indicators of poverty still predominate? *Development in Practice*; 17 (1): 4–13.

Swift J. (1994) Dynamic ecological systems and the administration of pastoral development. In: (I. Scoones, ed) *Living with Uncertainty: New Directions in Pastoral Development in Africa.* Intermediate Technology Publications; London: pp.153–73.

Sylvester C. (1999) Development studies and postcolonial studies: disparate tales of the 'Third World.' *Third World Quarterly*; 20 (4): 703–21.

Tamas P. (2004) Misrecognitions and missed opportunities: post-structuralism and the practice of development. *Third World Quarterly*; 25 (4): 649–60.

Taylor C. (1991) *The Malaise of Modernity.* House of Anansi Press; Toronto, ON.

Taylor C. (2007) *A Secular Age.* The Belknap Press of Harvard University Press; Cambridge, MA.

Thomas A. (2000) 'Meanings and Views of Development,' in T. Allen and A. Thomas (eds) *Poverty and Development: Into the Twenty-First Century.* Oxford University Press for the Open University; Oxford: pp. 23–50.

Thompson E. P. (1963) *The Making of the English Working Class.* Penguin Press; New York.

Thompson E. P. (1971) The moral economy of the English crowd in the eighteenth century. *Past and Present*; 50: 76–136.

Tilly C. (1984) *Big Structures, Large Processes, Huge Comparisons.* Russell Sage Foundation; New York.

Tilly C. (1985) Retrieving European lives. In: (O. Zunz, ed) *Reliving the Past.* University of North Carolina Press; Chapel Hill, NC: pp. 11–52.

Tilly C. (1990) *Coercion, Capital and European States.* Blackwell Publishing; Oxford.

Toye J. (1993) *Dilemmas of Development: Reflections on the Counter-Revolution in Development Economics.* Blackwell Publishing; Oxford.

Toye J. (1995) The new institutional economics and its implications for development theory. In: (J. Harriss, J. Hunter and C. M. Lewis, eds) (1995) *The New Institutional Economics and Third World Development.* Routledge; London: pp. 49–70.

UNDP (2000) *Human Development Report 2000: Human Rights and Human Development.* United Nations Development Programme; New York.

UNDP (2004) *Human Development Report 2004: Cultural Liberty in Today's Diverse World.* United Nations Development Programme; New York.

Uphoff N., Wickramasinghe M. L. and Wijayaratna C. M. (1990) Optimum participation in irrigation management: issues and evidence from Sri Lanka. *Human Organization*; 49 (1): 26–40.

Van Ufford P. Q. and Giri A. K. (2003) (eds) *A Moral Critique of Development.* Routledge; London and New York.

Vandergeest P. (2003) Racialization and citizenship in Thai forest politics. *Society and Natural Resources*; 16 (1): 19–37.

Vandergeest P. and Buttel F. H. (1988) Marx, Weber and development sociology: beyond the impasse. *World Development*; 16 (6): 683–95.

Vandergeest P. and Peluso N. L. (1995) Territorialization and state power in Thailand. *Theory and Society*; 24 (3): 385–426.

von Hayek F. (1944 [1994]) *The Road to Serfdom.* University of Chicago Press; Chicago.

Wade R. (1985) 'The Market for Public Office: Why the Indian State Is Not Better at Development,' *World Development*; 13(4): pp. 469–97.

Wade R. (1988) *Village Republics: Economic Conditions for Collective Action in South India.* Institute for Contemporary Studies; San Francisco, CA.

Wade R. (1990) *Governing the Market: Economic Theory and the Role of Government in East Asian Industrialization.* Princeton University Press; Princeton, NJ.

Wade R. (1996) Japan, the World Bank, and the art of paradigm maintenance: the East Asian miracle in political perspective. *New Left Review*; 217 (May/June): 3–36.

Wade R. (2001) Making the World Development Report 2000: attacking poverty. *World Development;* 29 (8): 1435–41.

Wade R. (2006) Choking the South. *New Left Review;* (38) March/April: 115–27.

Walker A. (2000) The 'Karen Consensus': Ethnic politics and resource-use legitimacy in northern Thailand. *Asian Ethnicity;* 2 (2): 145–62.

Wallerstein I. (1974) *The Modern World-System.* Academic Press; New York.

Wallerstein I. (1979) *The Capitalist World-Economy: Essays.* Cambridge University Press; Cambridge.

Wallerstein I. (1979 [2000]) The rise and future demise of the world capitalist system: concepts for comparative analysis. In: (J. T. Roberts and A. Hite, eds) *From Modernization to Globalization.* Blackwell Publishing; Oxford: pp. 190–209.

Wallerstein I. (1986) Review of *Big Structures, Large Processes, Huge Comparisons*, by Charles Tilly. *The American Journal of Sociology;* 91 (4): 981–4.

Warren B. (1980) *Imperialism: Pioneer of Capitalism.* New Left Books; London.

Watts M. (1983) *Silent Violence: Food, Famine, and Peasantry in Northern Nigeria.* University of California Press, Berkeley, CA.

Watts M. (2003) Development and Governmentality. *Singapore Journal of Tropical Geography;* 24 (1): 6–34.

Weber, M (1958) *The Protestant Ethic and the Spirit of Capitalism*, translated by T. Parsons. New York: Charles Scrinber's Sons.

Weiss L. (1998) *The Myth of the Powerless State.* Cornell University Press; Ithaca, NY.

White T. A. and Runge C. F. (1995) The emergence and evolution of collective action: lessons from watershed management. *World Development;* 23 (10): 1683–98.

Wiarda H. J. (2004) Introduction: the Western tradition and its export to the non-West. In: (H. J. Wiarda, ed) *Non-Western Theories of Development: Regional Norms versus Global Trends.* Thomson Publishing; Toronto, ON: pp. 1–19.

Willis P. (1977) *Learning to Labour: How Working Class Kids Get Working Class Jobs.* Saxon House; Farnborough, UK.

Wolf E. (1969) *Peasant Wars of the Twentieth Century.* Faber; London.

Wolf E. (1982 [1997]) *Europe and the People without History.* University of California Press; Berkeley, CA.

Wood E. M. and Foster J. B. (eds) (1997) *In Defense of History: Marxism and the Postmodern Agenda.* Monthly Review Press; New York.

Woodhouse P. and Chimhowu A. (2005) Development studies, nature and natural resources: changing narratives and discursive practices. In: (U. Kothari, ed) *A Radical History of Development Studies: Individuals, Institutions and Ideologies.* Zed Books; London and New York: pp. 180–99.

World Bank (1993) *The East Asian Miracle: Economic Growth and Public Policy.* Oxford University Press; Oxford.

World Bank (1998) *World Development Report 1998 Knowledge for Development.* Oxford University Press; Oxford.

World Bank (2000) *World Development Report 2000 Attacking Poverty.* Oxford University Press; Oxford.

World Bank (2004) *World Development Report 2004 Making Services Work for the Poor.* Oxford University Press; Oxford.

Wright R. (2004) *A Short History of Progress.* House of Anansi Press; Toronto, ON.

Ziai A. (2004) The ambivalence of post-development: between reactionary populism and radical democracy. *Third World Quarterly;* 25 (6): 1045–60.

Index

For Product Safety Concerns and Information please contact our EU representative GPSR@taylorandfrancis.com Taylor & Francis Verlag GmbH, Kaufingerstraße 24, 80331 München, Germany

Batch number: 08158431

Printed by Printforce, the Netherlands